Advance Praise for *Climate-Wise Landscaping*

Sue Reed and Ginny Stibolt bring complementary backgrounds to bear on the subject of how we can adapt our landscapes to a changing climate. Starting from the premise that the impacts of climate change will only become more severe in the future, the authors have created a comprehensive book that outlines dozens of actions that people can take to adjust to evolving climate regimes. In the process, they articulate a new gardening aesthetic for people who work with small garden plots, farms and woodlots. The result is a positive and hopeful story of how people can use their imagination and ingenuity to help craft more resilient landscapes.

— Dr. Peter Robinson, former CEO, David Suzuki Foundation

Climate-Wise Landscaping is a comprehensive, yet easy to read, source of information on climate-change adaptation and mitigation actions for the homeowner, gardener, and landscape professional. Beautiful photos and pleasing graphics illustrate key ideas and actions while informative sidebars and inspiring quotes from climate and landscape experts provide clarity of compl systems and motivation to adapt to a changing future. The text provides engaging blend of broad ideas along with specific actions we can take to ada to climate-change at the level of our home landscapes, whether a small garde plot within an urban area or twenty acres in the countryside.

— Julie Richburg, Ph.D., Ecologist

Given the lateness of the hour, a book on climate-wise landscaping could not be more timely or more necessary. We are moving into a new and critical era, and this book takes landscape professionals and home gardeners where they need to go. The facts, well-presented and practical, will be an eye-opener for many people, and empower us as horticulturists of all stripes to do what is both helpful and imperative. Landscaping has always been more than just exterior decorating, and now we have one more tool in our kit to make the landscape ecosystem a part of the solution. Thank you, Sue and Ginny, for hitting a home run.

— Owen Dell RLA, ASLA, landscape architect,
educator, author, Owen Dell & Associates

Gardening is not always as green and good for the planet as we might think. Sue Reed and Ginny Stibolt open our eyes to surprisingly common, unsustainable landscaping practices and inspire us to rethink how we create and care for land. This fantastic resource is filled with climate-wise solutions for anyone who owns or manages a piece of ground—even if it's just a few containers on a tiny rooftop garden. You will quickly learn about exiting ways to offset some of the effects we people have on the planet. The book is easy to navigate and it passionately links better gardening practices with better life quality and a brighter future for our planet.

— Claudia West ASLA, Principal, Phyto Studio LLC

The authors of this book provide a comprehensive approach to designing landscapes that have the potential of being both environmentally sound and experientially rich. I applaud them in doing the extensive research, backed up by their own professional expertise and experience, in creating this very useful guide to designing landscapes for the 21st Century which clearly respond to our changing climate.

— Darrel Morrison, Honorary Associate Faculty Member in
Landscape Architecture University of Wisconsin-Madison

CLIMATE-WISE
LANDSCAPING

Practical Actions for
a Sustainable Future

SUE REED AND GINNY STIBOLT

new society
PUBLISHERS

Cover design by Diane McIntosh.
Cover images © iStock: Garden 584877834; Trillium (bottom) 516010372.
Interior images: p 6 © Oksancia; p 10 background © alenek; p 228 © zgurski1980;
tree icons © topor / Adobe Stock

Printed in Canada. First printing March 2018.

Inquiries regarding requests to reprint all or part of *Climate-Wise Landscaping* should be addressed to New Society Publishers at the address below.
To order directly from the publishers, please call toll-free (North America) 1-800-567-6772, or order online at www.newsociety.com

Any other inquiries can be directed by mail to:

New Society Publishers
P.O. Box 189, Gabriola Island, BC V0R 1X0, Canada
(250) 247-9737

LIBRARY AND ARCHIVES CANADA CATALOGUING IN PUBLICATION

Reed, Sue, 1953-, author
Climate-wise landscaping : practical actions for a sustainable
future / Sue Reed and Ginny Stibolt.

Includes bibliographical references and index.
Issued in print and electronic formats.
ISBN 978-0-86571-888-3 (softcover).—ISBN 978-1-55092-680-4 (PDF).—
ISBN 978-1-77142-275-8 (EPUB).

1. Landscape gardening—Environmental aspects. 2. Landscaping industry—Environmental aspects. 3. Sustainable living. I. Stibolt, Ginny, author II. Title.

SB473.R43 2018   712   C2017-907028-2
                                                 C2017-907029-0

Funded by the Government of Canada · Financé par le gouvernement du Canada

New Society Publishers' mission is to publish books that contribute in fundamental ways to building an ecologically sustainable and just society, and to do so with the least possible impact on the environment, in a manner that models this vision.

new society PUBLISHERS · Certified B Corporation · FSC MIX Paper from responsible sources FSC® C016245

Contents

Acknowledgments

A Note from Author Sue Reed

Like many creative works, this book took a long time to produce and an even longer time to figure out before the writing began. I am very grateful for the help of many friends and colleagues throughout the entire process. My great appreciation goes to:

- Ginny Stibolt, for agreeing to share in creating this book, and for her gardening and botany expertise, excellent photos, clear thinking, and straightforward communication.
- Doug Tallamy, for his early validating comments about my idea for this book.
- Owen Dell, for his unique perspective on sustainable design, and for encouragement at both the start and end of this writing.
- Graduate students at the Conway School, class of 2016, for our fruitful brainstorming session.
- Sally Naser, for countless relevant articles and ideas, and always cheering me on.
- Bill Lattrell and Julie Richburg, for essential information about ecology and wildlife.
- Lauren Wheeler and Aaron Schlechter, for guidance about green infrastructure.
- Claudia Thompson, Katherine MacColl, Jan Voorhis, and Carol Cone, for sharing their gardening experiences. (And Katherine for editing help at the finish line.)
- Betsy Abert, for always reminding me about birds and their habitat needs.
- Adam Martin, for answering my many questions (at the end of one long hot day!) about community-scale composting.

- Benjamin Vogt, for sharing portions of his unpublished manuscript for *A New Garden Ethic*.
- John Cornell, for ideas about landscape construction and green roofs, and Fiona and Becky Cornell for sending me garden photos.
- Carole S. Brown, for including me as a contributor in her inspiring native plant blog.
- Rhea Banker, for her publishing advice at the start, many clarifying questions as I went along, and invaluable guidance about cover design at the end.
- Laura MacKay, for vital editing and marketing ideas.
- Marian Kelner, for thoughtful feedback about the whole manuscript.
- Ruth Parnall, for generous help with ideas and photos.
- Don Walker, for attentively reading a large portion of the manuscript and making it better; and for showing me the path of ecological landscape design 30 years ago, and steadily pointing the way ever since.
- Emma Picardi, for her gentle encouragement and constant confidence in my work.
- David Schochet, for his thousands of hours of intelligent listening, thoughtful editing, and clear advice, for improving the whole book, and for his unwavering, loving support.

Finally, Sue and Ginny give Marjorie Shropshire great thanks for her lively illustrations.

Foreword

by Doug Tallamy

Climate change is a global threat that is predicted to become one of the major drivers of the sixth great extinction life has suffered in its history on planet earth. But it is not the only driver of the loss of life from our landscapes. Though it gets little press, the way we humans landscape the spaces we have taken from the natural world has fragmented, isolated, and shrunken plant and animal populations wherever we have gone. We have accepted without thinking that humans and nature are like oil and water: they cannot mix. What is simultaneously frustrating and hopeful is that it doesn't have to be that way. There is no irrefutable reason why we can't live harmoniously with nature in the same place. Perhaps the need to address climate change with every tool at our disposal and with all of the ingenuity we can muster will be the stimulus for this cultural acceptance of nature everywhere. Fortunately, it is only a matter of will. As Sue Reed and Ginny Stibolt show us in *Climate-Wise Landscaping*, we already have the knowledge and we now have the desperate need to do this.

In this book you will learn that there are several basic roles each of our residential, corporate, municipal, and recreational landscapes must play if we are to fight climate change successfully. You will learn how and why we must build landscapes with plants that support food webs. Only such landscapes are capable of sustaining the biodiversity that runs the ecosystems that produce our life support systems, those natural products we call ecosystem services.

You will learn that to sustain these plants into the foreseeable future, we must sustain the pollinators on which indigenous plants depend, and we must do it on our built landscapes. There simply isn't enough Nature left to blithely pass this responsibility off to Her. Eighty percent of all plants and 90 percent of all flowering plants are pollinated by animals, and many of those plant/pollinator relationships are highly specialized. People often justify helping pollinators by their importance to our agriculture, but were we to lose pollinators, we would lose most of the plants on earth. This is obviously not an option if we wish to continue our residency on this planet.

Climate-Wise Landscaping will describe how to build sustainable landscapes that effectively manage our watersheds. It is plants that hold rain where it falls long enough for it to percolate through the soil and rock layers to be cleansed and to replenish the water table. It is plants that intercept the excess nitrogen and phosphorus we have put on our lawns and ag fields before it enters our waterways; and it is plants that soften the impact of pounding rain to help prevent the soil compaction that encourages rapid runoff. The plants we choose for our landscapes also create the biological corridors required to connect natural areas with each other. If we connect the habitats we have fragmented and isolated from each other, they will no longer be small and isolated, and that alone will slow the rate at which species are disappearing. After all, landscapes that do not sustain life can hardly be considered "sustainable."

Last but certainly not least, this book reveals the vital role plants of all sizes and shapes play in removing the excess carbon we have put into the atmosphere that is fueling changes in our climate; changes that are occurring far faster than we and other animals can adapt to. Nearly a third of the carbon now in the atmosphere has come from removing the forests and prairies that once covered much of the earth. Because plants are built from carbon, we help pull carbon out of the air every time we put a plant in the ground, regardless of its size. Moreover, plants pump carbon into the soil around them throughout their lives. If we can put plants back where they belong, they can infuse our soils with unimaginable amounts of carbon. Soil scientists now tell us that the earth's soils can sequester up to seven times the total amount of carbon presently in the atmosphere!

It is the plants in our landscapes and how we use them that will accomplish all of these goals. Read this book carefully. Everything you need to know to help heal our relationship with planet earth and empower you to make a much-needed difference is within these pages.

—Doug Tallamy, author of *Bringing Nature Home*
& *The Living Landscape*

Preface

Climate-Wise Landscaping combines the knowledge of two authors whose backgrounds and training differ but who share the same values, concerns, and hopes for the future. Sue Reed and Ginny Stibolt have decades of experience in their fields, Sue as a landscape architect and designer, Ginny as a botanist, gardener, and naturalist.

Both are teachers and communicators; both are committed to protecting the environment. Each brings her own regional experience and perspective to the question of how to manage the changing realities of this time, and how best to use our gardens and yards to help solve the problem of climate change.

Finally, Ginny and Sue both accept the scientific consensus that climate change is real and happening now, and that it is caused by an excess of greenhouse gases in the atmosphere. They see this era of climate change not mainly in terms of uncertainty and concern, but also as a time of opportunity and hope for human ingenuity. And they agree that now is not the time to indulge in despair about the situation, but instead to get to work in our own landscapes, on our community lands, on the grounds of schools, churches, and municipalities, to implement new ideas and solutions that will make a real difference.

Sue Reed

"I am a Landscape Architect with 30 years of experience designing landscapes in rural western Massachusetts. In all of my projects, I try to create comfortable, livable, ecologically rich environments full of native plants and mini-ecosystems. My goal is always to make sure people's landscapes work right, feel good, look great, save energy, and fit harmoniously into the larger natural world.

"When I started my design career, I would tell new clients that I would be using mainly native plants in their landscapes. Although some people were indifferent and others were intrigued, most had no clear idea what the term native plant meant. Now, 30 years later, public awareness and expectations have shifted so much that the majority of my clients *ask me* to design with natives. Just in the span of my own career, I have witnessed this dramatic change, and this gives me great hope that soon many more people will see their landscapes as places that, in addition to looking good, can also do good, for the environment and the planet."

Ginny Stibolt

"I am a botanist, naturalist, writer, gardener, and advocate for Mother Nature. I live and work in Florida, one of the states most vulnerable to the effects of climate change. Luckily, my house sits at the lofty elevation of 35 feet above sea level! My state is so large it contains seven Plant Hardiness Zones, which creates quite a challenge for me when I write about gardening here.

"When I joined the Florida Native Plant Society in 2006, it changed my gardening from what I 'could' grow to what belongs here. It's exciting to see people understand that an authentic Florida landscape filled with beautiful Florida native plants can become part of 'The Real Florida.' And in the long run, landscapes that are authentic to their local regions will be more resilient in the face of a changing climate."

Introduction

This book is a tool for anyone who wants to be part of the solution to climate change. It answers this question: What we can do, right now, in the landscapes of our own backyards and communities?

Predictions about the effects of climate change range from mild to unsettling to dire. But we are already experiencing warmer winters, longer summers, more frequent heat waves, and more extreme rains. Proposed solutions, at turns contentious, expensive, and complex, can leave the average person at a loss for what to do, or wondering whether anything can be done at all.

On this question, *Climate-Wise Landscaping* takes an optimistic view. Instead of wringing our hands, we prefer to roll up our sleeves. We are committed to making climate change the top priority, and helping others do the same, as we all design, build, and manage the countless gardens and landscapes throughout North America.

In these pages, you will find hundreds of easy, practical *Actions* that achieve at least one of the following goals; most achieve two, and many satisfy all three:

Credit: SReed. Designer: Walter Cudnohufsky.

All across North America, our landscapes can be both climate-wise and beautiful.

• Shrink each landscape's carbon footprint.
• Adjust our practices to create gardens and yards better able to flourish in new, challenging, and unpredictable conditions.
• Assist other species as they adapt to a changing world.

Most of these *Actions* are modifications of what we're already doing, as part of a hobby we love. Many cost little or nothing, and quite a few will save us money either now or later.

Without question, large-scale solutions are needed if we are to curb society's carbon emissions. And in fact, innumerable countries, cities, corporations, and individuals have already switched to renewable energy, or are in the process of doing so. Human ingenuity is blossoming, creating innovative products and solutions. In many places, renewable energy now costs less than fossil-fuel options. Public transportation systems are expanding in cities all over the world. But the big solutions will not be enough. They must be buttressed with millions of smaller steps and actions—one yard, one park, and one landscape at a time. This book lays out a path for those countless small but collectively transformational steps.

What's In It For You?

Taking just some of the actions presented in this book will:
• Save energy, money, and time
• Shrink the carbon footprint of any landscape
• Create cleaner air and water
• Create greater success with landscaping
• Increase physical comfort during extreme temperatures
• Support birds, butterflies, pollinators, and other wildlife
• Create healthier, toxin-free landscapes

Who Is This Book For?

The ideas in this book are useful for:
• Hobby gardeners and all homeowners.
• Amateur and professional landscapers.
• Master Gardeners.
• Landscape designers and students.
• Landscape caretakers, municipal planners, and property managers.
• Urban residents who wish to "green" their own cities and create healthier outdoor environments for their own neighborhoods and regions.
• All who are concerned about climate change and want to contribute to the solution.

This book's information applies to every kind of landscape project:

- Creating new gardens
- Renovating or restoring an existing garden
- Cultivating more food
- Expanding a natural area within one's property
- Creating a green roof
- Building a new home
- Countless other endeavors we enjoy in our gardens and grounds

Finally, the ideas presented here will be useful across most of North America. The book focuses on the prevailing conditions in the temperate regions of this continent; special notes point out exceptions for regions where conditions differ substantially from the temperate norm.

What's In This Book?

As stated above, the central theme of *Climate-Wise Landscaping* is how, in the face of daunting climate change, all of us can take steps to help solve the problem. Toward that end, the *Actions* presented in this book aim to help us shrink our carbon footprint and hold more carbon in the plants and soils of our landscapes. And, as it turns out, many of these *Actions* that lessen our carbon impact will also simultaneously work to make our gardens and yards more resilient, adaptable, and hospitable to wildlife.

The book contains ten Sections. Each one focuses on a familiar component of landscaping/gardening with a view toward particular goals:

- I: Lawn—Reducing the climate impact of lawns and their high-carbon maintenance.
- II: Trees and Shrubs—Cooling the air, storing carbon, stabilizing soil, and providing habitat.
- III: Water—Dealing with the future potential for both too little and too much water.
- IV: Ecosystems—Increasing diversity and carbon storage, and supporting wildlife.
- V: Soil—Taking advantage of soil's potential for greater productivity and carbon sequestration.
- VI: Planning and Design—Designing the whole landscape for resilience, vitality, and comfort.

- VII: Herbaceous Plants—Shrinking the carbon footprint of traditional gardening practices.
- VIII: Urban Issues—Dealing with urban heat and promoting nature in cities.
- IX: Food—Producing food locally and reducing CO_2 emissions.
- X: Materials—Assessing the various climate footprints of common landscaping materials.

Within each Section, several numbered *Action Topics* present lists of *Actions*: practical, doable, "close to home" steps many of us can take. Throughout the book, cross-references point readers to related topics and *Actions*.

Please note that this is not a *how-to* book. Although we provide some general guidelines for achieving a particular end, our intention here is to present an array of options that can be implemented in many situations and many ways, as appropriate for each individual landscape and gardener. So this is more of a *what-to* than a *how-to* book, more like an à la carte menu than a collection of recipes.

In addition, although this book does contain hundreds of ideas, it does not and could not possibly include every single good idea for making our landscapes more climate-wise. The aim here is to inspire, inform, and encourage, and to invite everyone to bring his or her unique creativity and genius to solving the problem of climate change—all while working to create beautiful and sustainable landscapes.

Never have we so hurt and mistreated our common home as we have in the last 200 years...yet all is not lost. Human beings, while capable of the worst, are also capable of rising above themselves, choosing again what is good, and making a new start.

— Pope Francis, Encyclical on the environment, "Care for Our Common Home," May 24, 2015

A Note about Who Wrote What

Although the authors collaborated on every part of this book, each Section was written primarily by one of us. Our two "voices," each representing specialized knowledge and experience, can be heard in alternating Sections; Sue was the primary author for the Introduction, the Conclusion, and the even-numbered Sections; Ginny was the primary author for the odd-numbered Sections.

What Is a *Landscape*?

The word *landscape* seems to mean many different things to many people. So, right here at the start, we would like to clarify the definition of landscape as used in this book.

A landscape is not simply a snapshot frozen in time. Nor is it a production to be done and finished, nor a picture to be appreciated only for its appearance. Instead, every landscape represents an ongoing process of change, evolution, experimentation, and surprise over time.

A landscape's visual appeal, which is subjective in any case, is a thing to be appreciated from many perspectives and angles, to be experienced in many ways, at many times of the day, throughout all the season, for many years.

And, beyond being a place that satisfies our own human wishes and desires—for beauty, delight, pleasure, privacy, exercise, creativity, sustenance, relief, or spiritual inspiration—each individual landscape is also a part of the larger natural environment, a home to other creatures whose lives depend on its resources and qualities.

As such, a landscape, then, is a place where we can also make a difference and serve a larger purpose. In our landscapes we have the opportunity to enhance the health and vitality of our own property, our neighborhood or region, and even the planet.

> You can hear the rain on leaves; the burring of tree limbs as they rub together; the hiss of wind across a field of dried grass. But that is not listening to the plants of this green world grow. Listen again—not with your ears this time but with imagination. You may hear the wrenching of bark as it forms its patterns; the whirr of a pollen grain through the air; the report of a bursting seed; the tinkle of sap in the tubes of a tree trunk; the twang of red rays ricocheting from the petal of a cardinal flower; the muffled sounds of roots expanding with the power of dynamite. The landscape is a vast system constantly in action.
>
> —Rutherford Platt,
> *This Green World*, 1942

What Is *Landscaping*?

From our perspective, landscaping is not just about creating gardens. Nor is it just decorating with plants and making things pretty. And it's not, as the term *landscaping company* might suggest, mainly about mowing lawns and keeping things tidy. No. Instead, we consider landscaping to be the process of creating experiences, including aesthetic experiences, that support our lives and also, where possible, the lives of other people and other organisms.

A PRIMER ON...

Climate Change

This book is not going to tell you how many gigatons of CO_2 your landscape emits in a year, nor exactly how many carbon equivalents of greenhouse gases your actions could remove or reduce. That specificity would be impossible in a book such as this. What the book *will* give you is ideas and advice about hundreds of ways to shrink your landscape's overall carbon footprint and make your landscape into more of a carbon sink than it is now.

The subject of climate change involves terms that may be unfamiliar or confusing to many people. This primer explains many of those used throughout this book.

It All Begins with Carbon

Carbon is a chemical element, the core element for life on earth. It can exist in a pure—or nearly pure—form, as in diamonds and graphite. But perhaps its most important quality is its ability to combine with many other elements to form a vast assortment of molecules. Such carbon-based molecules, often called the building blocks of nature, make up most of the tissue of plants, animals, fungi, and bacteria—almost all of life as we know it.

When carbon combines with oxygen, it forms the gas carbon dioxide (CO_2). A molecule of CO_2 contains one atom of carbon and two atoms of oxygen. In discussions about sequestering carbon—in plants and soil—we are referring to carbon in its *solid* forms. When we talk about greenhouse gases, we mean CO_2.

Greenhouse Gases

Carbon is the main ingredient in all fossil fuels, which consist, after all, of countless decomposed plants that sequestered carbon in their tissues during the eons past. When we burn fossil fuels for energy, their sequestered carbon is released into the air as carbon dioxide. And when large amounts of CO_2 enter the atmosphere, as has been happening for the past 200 years, this causes the *greenhouse effect*, i.e., the physical cause of global warming.

The greenhouse effect occurs when the sun warms the Earth's surface and not enough of that heat can escape through the atmosphere to let the planet cool off. The reason heat can't escape is the presence of *greenhouse gas* (GHG) molecules, which intercept infrared light as it is radiating outward. As a result, the lower part of the atmosphere stays warmer than it would be otherwise.

Gases that are considered to be greenhouse gases include water vapor (H_2O), carbon dioxide (CO_2), methane (CH_4), nitrous oxide (N_2O), ozone (O_3), and chlorofluorocarbons.[1] While the first five of these are produced by natural processes, their concentrations have been unnaturally increased by human activities such as burning carbon-based fuels. The chlorofluorocarbons are totally man-made, the result of industrial processes, aerosols, and refrigerants.

Each gas is present in the atmosphere at a different concentration. In addition, each gas has a different capacity for absorbing solar radiation. Methane is considered a much more potent greenhouse gas (in its atmosphere-warming capability) than carbon dioxide, but it lasts only about a decade in the atmosphere, whereas CO_2 will affect the climate for thousands of years.

Globally, the primary sources of greenhouse gases are:

- Electrical power plants: 25–30% (CO_2)
- Residential buildings: 11% (CO_2)
- Road transportation: 11% (CO_2)
- Deforestation and land use change: 10% (CO_2, nitrous oxide, and methane)

Smaller percentages come from these sources: energy industry processes (CO_2 and methane); commercial buildings (CO_2); cement and glass production (CO_2); livestock (methane); agricultural soil (CO_2 and nitrous oxide); waste and waste water (CO_2 and methane); coal mining (methane, CO_2); and aviation (CO_2, water vapor, nitrous oxide, and aerosols).[2]

Some Basic Facts about Climate Change

It was 1896 when the Swedish scientist Svante Arrhenius first described the existence of the greenhouse effect; it has taken us over 120 years to understand the dangerous implications of global warming, and that it is real and happening in the present, not in some distant future. How do we know?

- For the past 400,000 years, the planet's atmospheric carbon dioxide fluctuated wildly, mainly as a result of small variations in the Earth's orbit. This fluctuation caused cycles of extreme warming and cooling, but CO_2 levels in the atmosphere never exceeded 300 parts per million. In 1950, CO_2 levels rose above 300 parts per million, and they now stand at over 400 ppm.[3]
- The planet's surface temperature has risen about 2°F (1.1°C) since the late 1880s. Sixteen of the 17 warmest years on record have occurred since 2001.[4]
- Global sea levels rose about eight inches in the past century, with the rate in the last two decades nearly double that of the last century.[5]
- Since 1950, the number of record high-temperature events in the US has increased, while the record lows have decreased.[6]

- Satellite images reveal that the amount of spring snow cover in the Northern Hemisphere has decreased in the past 50 years, and that snow is melting earlier.[7]
- The acidity of the oceans has increased by about 30% since the start of the Industrial Revolution. This is the result of more CO_2 in the atmosphere being absorbed by ocean waters.[8]

A Few More Definitions

Gigaton (Gt) is an amount equal to one billion tons. This is the term usually used to indicate the quantity of carbon on earth, or CO_2 in the atmosphere.

Carbon reservoirs (also *pools* or *stocks*) are the places on the planet where carbon is stored. Aside from the rock that makes up the Earth's mantle, oceans represent the largest reservoir (40,000 Gt), followed by fossil fuels (4,000–5,000 Gt), worldwide soils (2,000–3,000 Gt), the atmosphere (700–800 Gt), and all terrestrial organisms (600–650 Gt). Surprisingly, the soil pool is larger than the atmosphere and all vegetation combined.[9]

The terms *carbon sink* and *carbon source* indicate processes of increasing or decreasing carbon amounts. A carbon *source* is a reservoir that is losing (emitting) more carbon than it is accumulating. Carbon *sinks* are gaining more carbon than they are losing.

Plants are carbon sinks because they continually take CO_2 from the atmosphere and convert it, via photosynthesis, into carbon (sugars) held within their biomass until they die and begin to rot. Soil is a carbon sink when it absorbs more carbon from decomposing plant roots (and other processes past and present) than it loses to tillage and disturbance (which allow carbon to combine with oxygen, producing CO_2, which is released into the air). Forests can be either carbon sources or sinks, depending on many factors, including how they are managed. Agricultural soils have been carbon sources for many decades, with their continual tillage and erosion, but modern regenerative agriculture is working to reverse that trend. Today the world's oceans are the primary long-term sink for human-caused CO_2 emissions—about 2 gigatons net uptake of carbon annually, but the oceans may become less efficient as the absorption of CO_2 causes them to become more acidic.

Carbon sequestration is a natural or artificial process through which carbon dioxide is removed from the atmosphere and held in solid or liquid form. Sequestration simply means storage. In the context of climate change, its purpose is to permanently reduce the amount of CO_2 in the atmosphere in order to help to reduce further warming of the planet. The sequestration process can be geologic, which involves capturing CO_2 before it is released into the atmosphere, and injecting it into deep rock formations. Or it can be terrestrial, in which land management practices like no-till farming and wetland/grassland/

forest restoration help to remove CO_2 from the atmosphere. In this book, we focus on this second method: the sequestration of carbon in plants and soil.

Carbon footprint is the total amount of greenhouse gas emissions caused by an individual, product, organization, or event. In most cases, carbon footprint can only be estimated because we don't have enough data yet or understand processes well enough to make absolutely precise calculations.

Climate footprint derives from the idea of carbon footprint. A broader concept, this term signifies overall impact on the climate, and contains greenhouse gas emissions along with efforts to sequester carbon and help the planet in general. While carbon footprint might be difficult to quantify, climate footprint really can't be accurately measured at all, at least not at the present time. It's a new and still-evolving term for how our choices and actions affect climate change, whether positively or negatively.

Regional Effects of Climate Change

Although no one can exactly predict the effects of climate change, evidence already shows that in the coming decades most of North America will probably experience warmer temperatures on average all year long. In the US Northeast and Midwest, this may mean shorter, rainier winters, whereas in the Rockies, Southwest, and Northwest, there will probably be less snowfall and less snowpack, leading to drier spring and summer conditions. We can already see that the Great Plains seems to be getting wetter in the north but drier in the south. And, while the Southeast might be warming less dramatically than the rest of the continent, Alaska and northern Canada appear to be warming twice as fast.

A little bit of greenhouse gas is good thing—life thrives in its warmth. The question is: how much? On Venus the atmosphere is 97 percent carbon dioxide. As a result, it traps infrared radiation a hundred times more efficiently than the earth. The earth's atmosphere is mostly nitrogen and oxygen; currently only about .035 percent is carbon dioxide, hardly more than a trace. The worries about the greenhouse gas are actually worries about raising that figure from .035 percent to .055 or .06 percent, which is not very much. But plenty, it turns out, to make everything different.

—Bill McKibben[10]

LAWN I

Overview

We begin this book with the subject of lawns for one simple reason: in landscapes all across this continent, changing the way we think about and deal with our lawns might be the easiest and most significant step we can take to help the planet.

Let's look at the big picture. The whole idea of *lawn* makes a lot of sense. An expanse of lawn offers a comfortable surface on which to walk, relax, and play. Lawn provides a simple stage where we can showcase our homes and landscaping. Planting grass is affordable and relatively simple, and taking care of it requires little specialized knowledge. In addition, when grass is actively growing, its countless blades give off wonderful oxygenated air all around us. And, at least as originally envisioned, lawn can be a way for us to have a piece of nature up close to where we live.

However. In the 100 years since the first gas-powered mower was invented—which made the lawn an achievable dream for the masses—the North American lawn has developed a big problem. And in the 60 years since the American Garden Club firmly instructed homeowners about the civic responsibility embodied in their well-manicured lawns, the lawn has lost some of the glory its early proponents once imagined.

The problem isn't the idea of lawn. No. Rather, it's what we've *done with it* that causes the trouble. As a society, we have multiplied much that's bad about lawn and in the process lost a lot of the good.

Vibrant expanses of lawn, once full of violets, clover, and many other little flowers, are now expected to be flawless monocultures, like a green carpet just delivered from the factory. To achieve this condition, many among us water lawns relentlessly: one inch per week no matter the season or weather. Regular dousing with synthetic chemicals makes the grass less safe for children and pets, and for wildlife. (See "A Primer on Landscape Chemicals" on page 19.) The emissions of mowers, blowers, and trimmers dirty the air. Summer weekends are full of din and clatter. And for this we spend millions of dollars on fuel, and emit tons of planet-warming carbon dioxide into the air. As a stage for our homes, and perhaps also as a symbol of social status, lawns are often made extremely large. And now, instead of being an inviting swath of nature, many lawns are expensive, barren green deserts, devoid of the vitality they were supposed to bring to our lives.

We can change this scenario, starting today.

Lawns occupy more than 63,000 square miles of the continental US.[2] This is 40 million acres, an area about the size of Texas. If just a portion of that expanse were managed differently or converted to something else—vegetable gardens, orchards, meadows, wildlife gardens—we would save billions of gallons of fossil fuels, reduce our CO_2 emissions, save billions of gallons of precious water, and protect countless waterways from being contaminated by polluted runoff. Oh, and while we're at it, we would support vastly more beneficial insects, birds, butterflies, and pollinators.

Figure I-1: More and more empty green lawns are being allowed or encouraged to grow tall and revert to natural vibrancy.

Credit: SReed

In This Section

On this continent's enormous acreage of lawn, a lot of us can take a lot of little steps to make a big difference. And we can continue to have, if we want it, some amount of traditional and comfortable (but nontoxic) green grass in our lives. This Section explains how to move toward that goal. We give you the following *Action Topics* and include "A Primer on Landscape Chemicals."

ACTION TOPIC   I-1   Liberate the Lawn

Why This Matters

Liberating lawns from dependence on chemicals is an important first step toward making the whole landscape more climate-wise. *Note*: For definitions and explanations of the substances commonly used in lawn and landscape maintenance, please read "A Primer on Landscape Chemicals," below.

Lawn-care practices vary widely between regions, neighborhoods, and individual landowners, but typical high-maintenance regimens include some of the following actions—actions that sometimes may be at odds with efforts to reduce our negative impact on the environment and shrink our climate footprint:

- *Regular applications of insecticides.* This will kill insects and other microorganisms in the soil. Unfortunately, the great majority of these living things are not harmful to lawns or grasses but instead play important roles in the soil and wider local ecosystem. For example, most of North America's 4,000 or so native bee species (our most important pollinators) raise their young underground. Some species build up an immunity to lawn poisons, so lawn care services apply them more often and add different poisons in reaction.

> Someday we might look back with a curious nostalgia at the days when profligate homeowners wastefully sprayed their lawns with liquid gold to make the grass grow, just so they could then burn black gold to cut it down on the weekends. Our children and grandchildren will wonder why we were so dumb.
>
> —Michael Webber[1]

- *Regular applications of fungicides.* Fungi play an important role in a healthy soil ecosystem because they digest organic matter and work with plant roots to help them better absorb nutrients, which increases soil carbon. Without enough fungi, the dead matter in soil decays more slowly, which can lead to a buildup of thatch. (See Section V for more details on fungi in soils.)

- *Regular applications of broadleaf herbicides.* In order to maintain a grass-only monoculture, these poisons kill off other types of plants. As with the insects, some weeds build up a resistance to the herbicides, so a lawn service might apply different mixtures to kill the weeds. But this can affect desirable plants, such as clover, which enrich soil and feed pollinators.

- *Regular applications of synthetic fertilizers.* After reducing the natural nutrients in the soil with applications of insecticides and fungicides, lawn services typically apply synthetic fertilizers so the grass will grow and stay green. But the nutrients in fertilizer are not bound to organic matter in the soil, so they tend to leach out during rain or heavy irrigation, polluting the groundwater and/or nearby waterways. Then, because they wash away so quickly, even more fertilizers are required to feed the starving grass.

- *Seasonal over-seeding with grasses.* These grasses grow actively during the normal dormant period. In southern regions, cool-weather grass species are used to keep the lawn green during the winter. In more northern regions, warm-weather grasses may be used to keep the lawn green over

Credit: SReed

Figure I-2: Who wouldn't love drifts of dainty "Quaker lady" bluets, splashed across the lawn? These blooms on their tiny stems usually fade in a month or so, leaving plenty of space for lawn grass to grow.

Credit: GStibolt

Figure I-3: Automated irrigation systems can be made smarter by adding moisture sensors that shut off the system after a rainfall, and by making sure that the water only irrigates plants—not the street or other hardscape.

the summer. This then requires more energy and more effort for year-round mowing and year-round irrigation.

- *Over-irrigation* An estimated 30% of household water is used for outside irrigation, and most of that is used for the luxury of keeping lawns perpetually green instead of letting them go temporarily dormant (tan).

In the ultimate irony, these conventional practices stimulate more and more grass growth, which then necessitates more and more frequent mowing, making the lawn a *carbon source* instead of the *carbon sink* it could be if managed more sustainably.

Actions

● **Stop using all pesticides and synthetic fertilizers.** When lawn treatments stop, some initial bug or fungal attacks may occur, creating thin, bare, or brown patches. Eventually, other plants will fill in the spaces, but don't worry about them; just mow everything equally. You might even come to love the diversity!

It may take a full year or more for the poison and synthetic nutrient residues to subside to the point that the lawn and its underlying soil will begin hosting enough alternative plants to be as green as it was before. This process can be speeded up by the addition of clover or other regionally appropriate, mowable species in the brown spots. In the long run, some of the turfgrass may survive, but because it will be mixed with a good variety of other plants, it will no longer be vulnerable to pests. (See I-4 on page 32 for information on repairing lawns sustainably.)

- **Reduce irrigation frequency, or stop watering altogether.** In many areas, municipalities have (or may soon have) water-use restrictions, so it's a good idea be ready for the possibility of no irrigation at all. If some irrigation is allowed and needed during a drought, apply water infrequently, but deeply. Frequent, light watering encourages roots to come to the surface, which makes the grass less drought tolerant. Remember: if lawn grows more slowly, this saves on mowing. The most climate-wise action is not to irrigate at all. (See III-2 on page 80 for on irrigation options, and I-4 on page 32 for more on taking care of liberated lawns.)

    Note: Lawn that goes brown during dry times is not always dead. Instead, it may have entered dormancy, a natural stage of life in many grasses that evolved in regions characterized by occasional or frequent drought. Climate-wise lawns are allowed to go tan in summer or brown in the winter!

    Regional Note: The one exception to allowing for dormancy is for properties in fire-prone areas, where it's important to keep lawn area mowed and

---

Are There Alternatives to Toxic Lawn Care Chemicals?

Yes. Natural landscape maintenance programs can achieve a healthy, pest-free landscape using the latest scientific developments in organic agriculture and horticulture. For example, corn gluten is a natural pre-emergent weed killer and fertilizer. Lawns can be enriched naturally by spreading a thin layer of compost in the spring and fall. Also, natural lawn care practices will lead to a healthy soil that supports plants in the lawn so they that resist pests and disease.[3]

relatively green near buildings. Dormant or tall grass could become fuel for a wildfire. (See VI-7 on page 180 for more fire-wise strategies.)

> **Use Integrated Pest Management (IPM)** All pesticides involve some amount of risk. The lowest risk for you, your family, and the general ecosystem results from using no pesticides at all. *IPM techniques* encourage natural processes to help ecosystems stay in balance, and let natural predators do most of the work. When done correctly, integrated pest management involves:

- Encouraging the beneficial organisms, including birds, bats, frogs, toads, lizards, predatory insects, and parasitoid insects. In this way, nature itself, with its large arsenal of bio-weapons, aids in keeping pests under control.

"The IPM approach compels you to consider your landscape as part of the larger community ecosystem to manage responsibly. The impact of your gardening and pest management decisions often extends far beyond your property lines." [4]

How "Lawn Service" Chemicals Affect Soil Ecosystems

Imagine what happens when we apply a general fungicide to the soil. The intricate dance of fungi and plant roots, which is so important to the health of plants, is slowed down or halted. Plants roots may not die, but their growth and vigor may decline. In addition, any insecticides that are used will poison the grubs, mole crickets, or other pests that feed on turfgrass roots, but they will also kill beneficial insects and macrobes, which means that the toads and other insect-eaters won't have the prey they need to survive. Since macrobes are also important soil aerators, without them plants will have a harder time growing. And, most importantly for climate change, when the soil is poisoned, the population of carbon-based life forms is drastically reduced, so the soil will not sequester nearly as much carbon as it could have. (See Section V on page 129 for more about soil ecosystems.)

- Discouraging the pests in a timely fashion, i.e., by emptying standing water every three days to discourage mosquitoes or by pulling annual weeds before they set seed, using crop rotation in the vegetable beds, cleaning up diseased plant material and removing it from the landscape (do not compost it).
- Planting the right plants for the landscape, and caring for them so they will grow well, which will reduce their susceptibility to infestation by insects and pathogens.

Last Thoughts

Liberating lawns from artificial fertilizers, pesticides, and unsustainable irrigation will save money, but more importantly, it is the easiest and most significant earth-friendly and climate-friendly action that homeowners, communities, schools, businesses, and municipalities can take. Liberated lawns are likely to be more resilient in an era of climate change, and the underlying soil will be able to sequester more carbon than lawns treated with synthetic lawn chemicals.

To some people, the drawback of the natural system is that it does not appear to work as quickly as the synthetic system. A natural fertilizer applied today may not turn your grass green tomorrow. In time, though, the natural system...will function better and get your grass just as green. Think of it as building a living, breathing underground community. The beneficial soil organisms that are neglected and often killed in the synthetic system will protect your grass, purify the groundwater, and keep your property safe for pets, children, and you.

—Paul Tukey, *The Organic Lawn Care Manual*, 2007

A PRIMER ON...

Landscape Chemicals

This primer defines and describes the substances that are most commonly used in landscaping, and particularly in conventional "lawn care." Before we begin this discussion, note the following:

- Everything is a chemical. Even water is a chemical, with two hydrogen atoms and one oxygen atom forming H_2O. So when people say that their yard or food is chemical-free, that's a misstatement. They probably mean free from synthetic chemicals.
- For the purposes of this book, we define a substance as *organic* if it consists of materials that have been alive, and we consider *synthetic* those chemicals made from materials that have not been alive.

Fertilizers

Early in the process of becoming farmers, humans realized that applying manures to the soil made plants grow better. Fertilizers—organic or synthetic—add nutrients to enhance plant growth. Plants mainly need nitrogen (N), phosphorous (P), and potassium (K), along with myriad micronutrients, which a healthy soil will provide. But if soil has been depleted, fertilizers must supply all these nutrients. The plants don't distinguished between synthetic and organic sources, but the *soil* reacts quite differently to each.

Synthetic Fertilizers

The manufacture of synthetic fertilizers has a huge carbon footprint: for every ton of fertilizer produced, two tons of CO_2 are released.[5] In addition, when too much fertilizer is applied, microbes in the soil emit nitrous oxide (a greenhouse gas) exponentially in relation to the amount of excess.[6]

"The use of artificial manure, particularly [synthetic nitrogen]...does untold harm. The presence of nitrogen in an easily usable form stimulates the growth of fungi and other organisms, which, in the search for organic matter needed for energy and building up microbial tissue, use up first the reserve of soil humus and then the more resistant organic matter which cements soil particles. In other words, synthetic nitrogen degrades soil."[7]

Organic Fertilizers

Organic fertilizers, such as compost and fish or seaweed emulsions, differ from synthetic ones because the nutrients are attached to complex structures, which add to the soil's humus (i.e., the organic component of soil, formed by the decomposition of leaves and other plant material by soil microorganisms). As a result, they don't rinse away as easily as synthetic nutrients, and they don't over-stimulate soil microbes to consume the available organic (carbon-holding) matter in soil.

Pesticides

Pesticides are substances designed to kill unwanted living things. These things may be plants, animals, fungi, and even bacteria. Some pesticides target only one type of organism; for example, Bt (*Bacillus thuringiensis*) targets only caterpillars. However, most pesticides, whether organic or not, kill a wide spectrum of organisms. *Pyrethrins*, derived from chrysanthemum flowers, are organic, but they are still highly toxic to many life forms.

In addition to their mostly negative effects on the local ecosystem (see below),

The Poison Cycle

A balanced ecosystem contains many prey organisms—aphids, whiteflies, mosquitoes, cabbage worms, leafminers, mole crickets, spider mites, etc.—which provide food for a corresponding number of predatory insects, toads, bats, frogs, and birds, which together keep the ecosystem in balance.

When poisons (organic or synthetic) are used on "pests," the majority of all the insects in an area will be killed. But more than 90% of those insects are beneficial or benign. Some of those beneficial insects would have pollinated flowers, while others would have eaten some of the unwanted insects. Other predators such as bats, frogs, and birds will go else-where to feed, so the poisoned property will have lost its valuable predators.

As a landscape recovers from the poison, unwanted insects will begin to multiply again, but since their predators were killed or went away, the harmful insects will recover in even greater numbers than before. When sprayed again, the process repeats itself, and each time the most damaging insects will recover in ever-increasing numbers. Repeated poisonings often foster resistance to that pesticide. It would be better, for our own health and the health of the environment, to break that cycle and manage the landscape as a complete ecosystem where prey and predators are balanced.

the carbon footprint of pesticide manufacture and transportation is large. Several types of pesticides are described below.

Insecticides

Despite the fact that the vast majority of insects in a typical landscape are either beneficial or benign, landscape-wide, broad-spectrum insecticides are regularly applied—usually as part of traditional lawn care, to kill the few insect larvae (grubs) that chew on grass roots.

Neonicotinoids

There are various types of systemic insecticides that are absorbed into the entire plant. Those called *neonicotinoids* are widely used in the flower-growing trade to keep the plants looking good (uneaten) on the shelves of garden centers. These substances target the nerve impulses of sap-sucking and leaf-chewing insects and other invertebrates, and so are thought (but not yet conclusively proven) to be safe for humans and other mammals. However, when these substances are incorporated into pollen and nectar, many beneficial pollinator insects (including thousands of native bee species) are harmed by these poisons.

Homemade Insecticides

There are concoctions (mostly mixtures of soaps or oils) that can be used on insect infestations such as aphids, but if a substance kills aphids, it will also kill pred-

atory insects like ladybug larvae. Soaps also dissolve the plant's waxy cuticle that protects it from insect and fungal attacks. Just because it's homemade doesn't mean that it's not toxic.

Fungicides

Fungicides are often included as part of a routine lawn-care regimen and are used preventively for fruit trees. In most landscapes, regular use of fungicides generally causes more harm than good because, as will be further discussed in this section and in Section V, fungi play vital roles in a healthy, balanced soil ecosystem. They help plants survive by aiding in the absorption of water and breaking down organic matter in the soil. In addition, fungi can help to limit soil pathogens. In climate-wise landscapes, it's preferable to keep and encourage soil fungi.

Herbicides

Herbicides kill plants. *Pre-emergent* herbicides suppress seed germination, and *post-emergent* herbicides target existing plants/weeds. In this latter group are two types:

Broad-spectrum Weed Killers

There are a number of broad-spectrum weed killers on the market, but glyphosate-based products are most commonly used. Widespread application of these products, particularly for crop management (on both GMO and non-GMO crops such as wheat) can have far-reaching effects that we are only now beginning to fully understand. There is mounting evidence that traces of glyphosate are in our food. (A side effect of their use is that milkweed, which once was plentiful around crop fields, is now mostly missing, and, as a result, the numbers of migrating monarch butterflies have been depleted.)

Note: Smaller, non-farm landscapes might not have these problems, but let's work to minimize broad-spectrum herbicides and their unknown side effects. One strategy, instead of spraying widely, is to cut back any persistent weeds to the ground; when they grow back, apply the herbicide just on that new growth—when the plant is most vulnerable.

Broadleaved Weed Killers

2,4-D (2,4-Dichlorophenoxyacetic acid) is probably the most-used weed killer for lawn-care regimens, but in the 15 days or so that it is active, it can leak into groundwater and nearby waterways, where certain forms of it are toxic to fish.[8] Other products such as Dicamba and Triclopyr are also used in lawns. These can sometimes be absorbed by tree roots growing under lawns, harming the trees themselves.[9] Ironically, the use of broadleaved weed killers also weakens the grass, making it more susceptible to fungal and insect infestations. The widespread practice of creating monoculture lawns, which can only be supported by chemical treatments that then stimulate the need for more chemicals, can be an expensive and self-defeating cycle for homeowners.

Homemade Weed Killers

Most recipes for homemade weed killers include vinegar, salt, and/or soap. Yes, these substances will sometimes kill plants given the proper conditions, but vinegar sprayed into a bunch of weeds will also kill toads that were hiding there, and it will leach into the soil where it might also kill worms and other soil inhabitants as it acidifies the soil. With any homemade herbicide, be careful to spray only the plants.

A Cautionary Note about Herbicides

Although herbicides can be effective in eliminating unwanted vegetation, they also pose multiple threats to the environment if used improperly. Plus, even if used properly, herbicides may affect soil microorganisms and the functioning of soil ecosystems, which will reduce the soil's ability to sequester carbon.

Figure I-4: This sign says it all.

ACTION TOPIC (I-2) **Reduce the Extent of Lawn** »

Why This Matters

Lawns have become the default landscape feature in almost every region of the continent, in every type of neighborhood, and for every kind of house. But removing lawns or reducing their size will save water, reduce pollution (including greenhouse gas emissions), and better support wildlife. In addition, smaller lawns mean less mowing, which by definition leads to a reduction in CO_2 emissions.

Credit: SReed

Actions

Before taking any action, create a plan for other vegetation or landscaping to replace the lawn. Don't leave it to chance. Options might include:

- Creating a grass/wildflower meadow (see I-7).
- Building a rain garden (see III-4).
- Planting a productive vegetable or herb garden, or growing berry bushes and fruit trees (see Section IX).
- Creating a habitat garden (which might be just a meadow enhanced with wildlife-supporting features such as logs on the ground, standing snags, birdhouses, and birdbaths (see Section IV).

Figure I-5: Reduce the size of the lawn and mow less frequently to reduce the carbon footprint of any landscape.

Carbon Footprint of a Lawn Mower

Here's one way to calculate the carbon footprint of just mowing lawn: "1 hour × 22 mowings × 0.5 gallons of gas for each mowing = 11 gallons used per year. 11 gallons of gas × 17.7 pounds of carbon per gallon = 194 pounds of carbon emitted per year."[10] This footprint is even larger when edging and blowing are included. (See more about this in I-3 on page 28.)

- Planting a grove of trees or hedges and shrubs to help shade the ground (or the house) and to provide important vertical layers and diversity in the landscape (see IV-3).
- Constructing a patio or outdoor sitting/gathering space (see Section VI).

❯ **Remove the lawn**, either all at once or a little at a time. For some situations, removing a whole lawn and then replanting the area with groups of native plants and/or edible gardens in one large project may be the best choice. This usually works for municipalities or communities that have the funds to buy the plants and hire a workforce, or that can muster groups of community service volunteers for workdays. More often for homeowners, lawns are reduced in stages depending upon budget and available time.

When doing the lawn removal in stages, start with a plan for removing lawn from the edges of the lot, under trees, in wet or low spots, along slopes, and in ditches. Look for areas where the grass is not doing well and mark those areas for grass removal. Then plan for new vegetation that merges with or complements the adjacent non-lawn area, whether it is a garden bed, a meadow, or a wooded area. Taking on one area at a time has the advantage of allowing you to find out what works and what doesn't—some plants will thrive, while others will struggle or fail. These lessons can be applied for the subsequent lawn removal stages.

Two possible scenarios where replacing lawn makes sense:

In a Wet Spot:

If a mowed area is habitually damp, mowing it is likely to be difficult due to wet or muddy conditions. It's also possible that rushes, sedges, or other moisture-loving plants are already growing among the lawn grasses. Simply stopping mowing would release these species to show their full character. De-lawning here could include planting a good variety of plants that are appropriate for rain gardens—including trees and shrubs if there's enough space—in attractive groupings with a neat border. Almost any treatment that can survive in a wet area will be better than trying to maintain a lawn there, and the new vegetation could serve as habitat for wildlife, including important insect predators such as frogs, toads, and dragonflies. If a wet area occupies a back corner and is shared with neighboring properties, maybe everyone could work together to

Credit: SReed. Designer: Ruth Parnall.

Figure I-6: Front lawns can be completely lawn-free, like this one! (But be sure to check with local lawn ordinances, and help re-write them if necessary.)

create a unified feature that would enhance all the properties. (See Section III for ideas on rain gardens and water features.)

Where Shallow Tree Roots Have Out-competed the Turfgrass:

Many tree species such as maples have widespread and shallow root systems, which are not compatible with growing turfgrass. One approach might be to replace all the grass with a mostly mulched area inside the drip-line, especially when there are low-hanging branches. If there are narrow areas of

I always encourage my clients to practice intentional "lawn-i-cide," so they can conserve water, increase the habitat value of their garden, eliminate the use of gas-powered and polluting mowers, and escape the paralyzing boredom of an uninterrupted swath of green. Time to ignore the siren song of the "perfect lawn."

—Billy Goodnick,
Landscape Architect,
Santa Barbara, California

Methods for Removing Lawn

Some methods for removing lawn are more climate-wise than others; anything that minimizes soil disruption and doesn't burn fuels has a smaller carbon footprint. Options include:

- *Stop mowing, fertilizing, and watering.* In some drier climates, the turfgrass will die on its own when all its artificial support is removed. In more diverse lawns, some plants may survive and (unless they are known invasives for the region) you should let them grow. Add other native meadow plants such as attractive bunching grasses and

Credit: SReed

Figure I-7: Tree roots are often incompatible with lawn: here is a perfect opportunity to create a wildflower meadow, grove, or mini-ecosystem around a solitary tree.

drought-tolerant perennials or freely seeding wildflowers.

- *Smother the lawn.* Leaves, wood chips, or other readily available organic materials, laid over the lawn about six inches deep, will block the light, and, if left for six weeks or more, should kill most of the plants growing in the lawn. (Use a much thinner layer under trees so their roots are not damaged by lack of oxygen.) This treatment is climate-wise because there is a minimum disturbance of the soil, and the dead plants under the layers create another layer of mulch.

- *Solarize the grass.* Cover the lawn area to be removed with plastic sheets and weigh down the edges with bricks or rocks to keep the sheet in place for at least six weeks—summer is the best time of year for this treatment. After removing the plastic, add a thin layer of compost (about ½ inch) to revitalize the soil's ecosystem and help decompose the dead lawn.

- *Physically remove turf.* This can be done manually, using cultivators, shovels, and rakes; or mechanically, using a plow, rototiller, or specialized sod-cutting machine. Manual removal is recommended for areas up to 30 feet around established trees, so their roots are not harmed too much. If the sod is removed, it could be used in a compost pile where it could be layered with dead leaves or other brown materials to build a hot compost. (See more about compost in V-4 on page 144.) The advantage to this method is that it's quick—given the right tools and enough workers—and the replanting can begin right away. The disadvantage of this method is that soil disturbance will release CO_2, and the use of fossil-fuel-based tools will emit CO_2 as well.

- *Last Resort*: Apply herbicide. For some projects, particularly large ones, lawns can be eliminated with an application of a nonselective herbicide, like glyphosate. Before applying any herbicide, be sure to read and follow the instructions. (For more about herbicides, see "A Primer on Landscape Chemicals" on page 19.)

distressed grass jutting into the lawn from an adjoining wooded area, you could create a transition area from the trees to the lawn by planting a variety of shrubs and/or perennials between the larger tree roots. Make sure to create an-easy-to-mow edge.

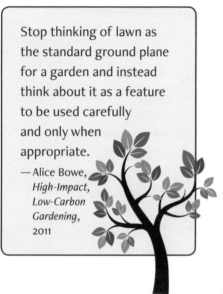

> Stop thinking of lawn as the standard ground plane for a garden and instead think about it as a feature to be used carefully and only when appropriate.
>
> —Alice Bowe,
> *High-Impact,*
> *Low-Carbon*
> *Gardening,*
> *2011*

Last Thoughts

Replacing lawn with better habitat will help birds and valuable pollinators manage the coming climate changes in local conditions. Growing edible gardens and orchards instead of lawn will save money and reduce the CO_2 emissions associated with commercial food production and transport (see Section IX). Creating rain gardens will help sequester stormwater, which reduces pollution of local waterways and the need for expanded municipal sewer systems; and rain gardens can even help replenish groundwater or aquifers, which are under strain. And, most important for creating a climate-wise landscape, replacing lawn with anything at all that doesn't require mowing, trimming, fertilizing, irrigating, and treating with chemicals will shrink the substantial carbon footprint of lawns and lawn maintenance. (See I-3 below for more about mowing less).

ACTION TOPIC (I-3) **Minimize Use of Power Tools** »

Why This Matters

According to the EPA, Americans burn about 800 million gallons of gas each year trimming their grassy yards. And they spill an additional *17 million gallons* just in refueling their equipment![11] So even though lawn grass does sequester some amount of carbon (the amount depends mainly on soil health; see Section V), this is generally outweighed by the amount of carbon dioxide emitted in làwn management.

In addition, small engines, especially older 2-cycle engines used for lawn care (including mowers, blowers, trimmers, leaf vacuums, and edgers), emit high levels of pollutants such as carbon monoxide, hydrocarbons, and nitrogen oxides. Relatively new emission controls for small gasoline engines have helped the situation to some degree, but this equipment still causes significant air and noise pollution.[12]

Actions

> **Reduce the size of lawns, or remove them.** Simply put: smaller lawns equal less mowing. (See I-2 on page 23 for more on this topic.)

> **Switch to manual, electric, solar, propane-powered, or battery-powered equipment.** Electric or solar-powered tools produce essentially no pollution from exhaust emissions or through fuel evaporation, and they are much quieter. Consider using solar or battery-powered robotic mowers. Equipment that runs on propane is cleaner running than gas-powered equipment. If possible, the best option is to use manual tools for small yards and small jobs. The estimated carbon footprint (carbon emissions associated with manufacture, distribution, sales, etc.) of a push mower is a *total* of about 32 pounds of carbon (compared to the gas mower footprint of 194 pounds *per year*).[13]

> **Mow less frequently; even allow lawn to go dormant.** Minimize use of fertilizers and use only organic, slow-release materials such as compost. Apply them only at the beginning of the natural growth season, as discussed in I-1. This keeps the soil healthy and maintains slow grass growth. Do not over-seed lawns with grass species that will grow in the off-season. Don't mow at all during dormancy.

Credit: GStibolt

Figure I-8: We can design landscapes to minimize maintenance, but if power equipment is needed, battery-powered tools are quiet and emit no fumes; interchangeable battery packs can work in several different tools.

How to Design Lawn Edges for Easy Maintenance

Reduce the need for string trimmers and edgers by incorporating better design, which could include these elements:

- Around the edges of garden beds, set edging timbers, bricks, or other pavers flat and just slightly higher than the lawn. This way, the mower's wheels can ride smoothly along the pavers and cut that edge grass easily, without needing additional trimming. This strategy might also work for areas under or along fences.
- Woodchips or other organic mulch can be used to define the edge of a lawn to separate it from non-lawn areas. Make sure that the mulch is tamped down to provide a firm surface for the mower's wheels.
- Plants in the beds abutting the lawn may need to be trimmed back periodically so they don't hang over the pavers or mulch and interfere with the mowing process. If they are growing well and you don't wish to prune them back, expand the bed to accommodate the plants and move the pavers or mulch out to a new edge. This is also a good way to gradually reduce the size of the lawn.
- Reduce the growth of grass runners that crawl over the pavers or mulch and into beds by bending them back and weaving them into the lawn area instead of cutting them. Cutting the runners will release sprouting hormones, which results in new growth spurts of the runners, but folding them back does not.
- Replace difficult-to-mow sharp corners with gentle curves that a lawnmower can easily negotiate. Broad, gentle undulations in the borders of the lawn provide interest in the overall landscape.

Figure I-9: Using pavers or bricks set just above the level of the lawn provides a solid pathway for the mower that allows it to cut the grass at the edges of the lawn. This eliminates the need for extra trimming tools.

Credit: MShropshire

Mow less often during the growing season and set the mower blade to the highest recommended level for your grass. This allows the grass enough leaf area to photosynthesize, but also to shade the soil, which helps to reduce weeds. As new species work their way into the turf, set the blade higher. Use a variety of mowing patterns so soil compression along the mower's tracks is evenly distributed. For maximum soil health and carbon sequestration, use a mulching mower to better chop up the blades of grass and other organic matter, and let this valuable material return through the turf into the soil.

> **Properly maintain power tools.** Follow the manufacturer's guidelines for maintenance by cleaning equipment often, keeping engines tuned, and keeping blade(s) sharpened for the most efficiency. Recycle old power tools where they will be disassembled for reuse of raw materials, so that they are taken out of service and will no longer pollute the environment.

> **Educate lawn service crews.** Establish service contracts so lawn companies practice more sustainable methods. At a minimum, this should include:
> - Minimal use of fossil-fuel powered equipment.
> - No pesticides or herbicides, and the absolute minimum of slow-release fertilizer.

Credit: SReed

Figure I-10: Lawns that are allowed to go dormant during dry spells will usually spring back to life as soon as moisture returns, unless a drought is extremely protracted.

"Until recently, emission controls for small gasoline engines were not a crucial design consideration for manufacturers. Consequently, small engines contribute more emissions per hour of use than most cars, because cars have complex emission control technologies installed. Power equipment users may also inadvertently contribute to pollution through careless fuel handling and improper maintenance."[14]

- Grass clippings retained on site, either as mulch within the lawn or in a compost pile.
- No mowing when the grass is dormant; for year-round contracts, work could focus on other tasks than mowing, i.e., removing areas of lawn from around trees or raking thatch (which should be minimal in any case).

Last Thoughts

For those who prefer to have some amount of lawn, reducing the use of mowers and power tools is one of the best ways to reduce a landscape's climate footprint.

ACTION TOPIC (I-4) **Revive Damaged Lawns** »

Why This Matters

When a lawn is liberated from a high-maintenance regimen, as described in I-1, the grass will probably go through a period of adjustment to the new normal. Remedial action might not be needed but, in general, a variety of actions will help the lawn and its underlying soil begin to function normally more quickly.

Note: This information applies to regions that receive enough natural rain to support a lawn, i.e., usually more than 20 inches of rainfall annually. Arid and semi-arid climates can support wonderful landscapes, but not turfgrass or mixed lawns.

Actions

Apply compost. Since the soil's ecosystem is likely to have been damaged by repeated applications of fungicides and other toxic substances, compost will invigorate the soil's microbes that may have been depleted. Compost also supplies some organic nutrients bound to humus, which remains in the soil longer than synthetic fertilizers. Compost also helps to reduce thatch in lawns by introducing microbes that will work to decompose the dead materials (see below). The compost should be made from local materials (fallen leaves, grass clippings, arborists' wood chips, and/or gardening trimmings, but no manures) (See V-1 on page 144 for more on building compost.)

Credit: GStibolt

Figure I-11: After treatments of pesticides and synthetic fertilizers are stopped, some parts of the lawn may turn brown. A thin layer of locally made compost (with no manures) applied to these distressed patches can help rejuvenate the microbes in the underlying soil.

Apply a thin layer of compost or compost mixed with sand, ½ inch or less, but only at the beginning of the turfgrass growing season. In the Southeast, apply it in the spring just as the grass is beginning to grow for the season and well before the summer wet season, which starts in June—and never in the fall when it's going into dormancy for the winter. In more northerly regions, the beginning of fall is the recommended time for compost. This process can be repeated on a yearly basis, but usually one application is enough to boost the health of the soil's ecosystem.

Note: Skip this soil-building step on land that is naturally low in nutrients, because lawns are not a sustainable option where soil is extremely poor. Instead, we suggest either enjoying what will grow there naturally, without fighting Mother Nature, or creating raised beds for special projects.

Credit: SReed

Figure I-12: Raking away thatch—dead lawn waste—opens up the soil surface so rain or irrigation can sink in more easily. For sustainably managed lawns, thatch is less of a problem than in chemically treated lawns because dead materials are quickly consumed by the organisms in healthy soil. It's a good idea to dethatch the lawn before adding compost.

Reduce Thatch. Thatch is the buildup of dead and living plant material above the soil and under the growing turf. It is generally the result of grass growing faster than the soil microbes can break down the dead plant parts, and it is usually caused by a combination of too much fertilizer, fungicides, shallow irrigation, soil compression, and root damage.

Thatch is more prevalent in the types of lawn grass that spread with aboveground runners, called *stolons*. An indication of thatch is when the turf feels spongy. When thatch is an inch thick or more, it repels water and anything else applied to the lawn. Also, because thatch elevates the growing parts of the grass, standard mowing may scalp the grass too close for optimum growth.

To reduce thatch, stop applying fungicide and cut back on both irrigation and fertilization, as recommended in I-1. Then, for a small lawn during its dormancy, vigorously rake the lawn area with a flexible leaf rake to pull out much of the dead plant materials (this material is a great addition to the compost pile because it's already partially rotted).

To fight thatch on a larger scale, use methods that also work to remedy compacted soils, as discussed in V-3.

Last Thoughts

The most sustainable way to prevent thatch buildup and other problems is by employing good lawn management practices as discussed in I-3 and I-4. However, taking an active role to help the lawn recover will increase both the rate of carbon storage and our own pleasure in whatever (modest?) amount of lawn is in our care.

C-edit: SReed

Figure I-13: This grass road goes dormant in summer because its soil is thin and highly porous (due to the gravel base needed to support vehicles), but year after year it comes back to life with no assistance.

Although each grass seed that germinates when the lawn is sown is an individual plant, the roots and stems soon form a felt-like mat from which the individual plants can no longer be untangled. The mat is seldom thicker than 4 inches; where developers have been stingy about replacing the topsoil they scraped away, the carpeting may be a mere 2 inches thick. Naturally, so thin a skin dries quickly in the summer sun; unnaturally, we don't allow it to protect itself by tanning and taking its normal midsummer nap. To keep our grass green all summer, [experts] prescribe 1 inch of water a week, a total of 24 inches during the growing season, or more water than falls as rain from May through October anywhere in the United States.

—Sara Stein, *Noah's Garden*, 1993

ACTION TOPIC **I-5** **Provide Good Preparation for New Lawns** ≫

Why This Matters

A new lawn established with good soil preparation and planted with a mixture of appropriate grasses, clovers, and other plants can serve as a useful feature in the landscape with a minimum amount of ongoing care. This information is applicable in temperate regions that receive at least 20 inches of natural rainfall annually, sufficient to support a lawn.

Actions

❱ **Map out the minimum area needed for the lawn.** It's most climate-wise when lawns are designated to serve specific functions within the landscape, such as a play yard, a dog run, or as the floor of an outdoor room. As discussed in I-3, plan for any mulched beds, hedgerows, and other non-lawn features to have wide curving edges for easy mowing (which will reduce mowing time and associated fuel-consumption). If the lawn is small enough, the mowing might be easily accomplished with a manual reel mower or a small rechargeable battery-powered mower (ideally, using electricity supplied by on-site solar panels!).

❱ **Prepare the soil.** Before you seed, sod, or sprig (i.e., planting evenly spaced plugs of turfgrass) the new lawn, invest the time and resources into preparing the soil. Just plopping sod or seed on unprepared or poorly prepared soil may appear to be the faster, less expensive method, but it's a false economy in the long run.

Sustainable Lawn Is Possible Only in Non-arid Regions

Arid and semi-arid climates that receive less than 20 inches of annual rainfall are unlikely to support a lawn, even if managed sustainably. In arid regions, there are better—water-wise, climate-wise, and cost-wise—alternatives to lawn. Homeowners in arid Las Vegas have been paid to remove their lawns; maybe we can find some lessons there. Some arid regions do receive rainfall during different times of the year. In that case, observing local untended or natural areas can provide important information about what type of plant communities are supported. If prairies or other dense plant cover grows in untended areas, then landscapes in that region may be able to support turfgrass, but these lawns generally will be less sustainable than those where natural precipitation is greater.

In some cases, grading—with or without power equipment—may be necessary to create a smooth, even mowing surface; but you should work to minimize soil disturbance to help maintain the soil's ecosystem (see V-2 on page 138).

The best time to work on the soil is before installing the grass and other lawn plants. Begin the process by testing existing soil to see which grasses will do best there, and talk to the local extension agent for recommendations. Soil with plenty of humus will increase the durability of turfgrass and other lawn plants to withstand drought and other stresses. If the soil is sandy or clayey, add a layer of compost to increase the volume of organic matter, or humus. This will increase water retention in sandy soils, provide more air spaces in clayey soils, and create an inviting substrate for beneficial organisms. In a situation where there is bare soil, it's a good idea to mix the compost 50/50 with sand to keep it in place while the seeds or sprigs are getting established. The compost mixture should be laid on top of the soil and then raked in before planting or seeding. This way, the underlying soil is not disturbed as much as it would be if the compost were dug in. (For more information on compost, see V-4 on page 144.)

> **Provide sufficient depth of soil.** Bear in mind that the depth of good soil you

Credit: SReed

Figure I-14: After construction of this new home, the small lawn is prepared with topsoil saved and reused from the site demolition. An autumn seeding of "No-Mow" grass seed (a dwarf fescue blend) yields a dense cover of 5–8″ tall lawn, which can be mowed either occasionally or never, as desired.

provide has a direct and often quite apparent relationship to the health of the grass. Grass roots will only go as deep as they can find nutrients in the soil. Seeding lawn in soil shallower than 4 inches deep usually results in the grass going dormant (tan) during dry periods. A better solution is to provide soil that's 5 or 6 inches deep. This initial investment will repay you in easy maintenance and healthy lawn for decades to come.

Important note: Make sure that any soil to be seeded is not overly compacted. It should feel firm and slightly resilient, not hard but also not squishy. Soil that lacks air pockets will prevent the growth of new grass's tiny roots, inhibiting their ability to take up nutrients and water. Soil that's too loose will settle in unpredictable patterns, potentially creating voids in the soil where roots will dry out, or causing puddles on the surface where standing water could drown the grass. (See V-3 on page 142 for more on soil compaction.)

Choose the right plants and method of installation. Choose the mixture of grasses and other plants best suited to the soil, available irrigation, local climate, salinity, and predicted traffic. Some new turfgrass cultivars have recently been developed that require less fertilizer, less water, and have a high salt tolerance. Don't just accept what's available at a big box store; find the best choice for your regional conditions. In most locations, clover is a good addition to the lawn mixture because it's a legume and can provide its own nitrogen nutrients. A mix of native grass seeds could be added for more resilience. Over time, the grasses will probably be diluted with other species of plants when the lawn is maintained without poisons. As long as the new plants are not invasive in the region, are not thorny, and will tolerate mowing, they should be welcomed as part of the mix. Do some research with a local nursery, farmer's co-op, cooperative extension, or native plant society for advice about which types of grass and other ground covers will thrive best in your particular conditions.

Figure I-15: Sod is the fastest and probably least climate-friendly method of installing a lawn, but in cases where the landscape includes significant slope where soil could erode, it may be a reasonable option. If sod is used, good soil preparation will pay off in the long run.

Credit: GStibolt

Seeds are usually the most climate-friendly way to start a lawn, but sprigs, sprigs interspersed with seed, or sod may be reasonable options for some situations, especially sloped areas. Unless there is heavy precipitation, irrigate deeply every other day or so for two or three weeks until the seed has germinated or the sprigs or sod are established. Gradually reduce the irrigation until the lawn can survive on its own. If possible, time the establishment of a new lawn with the beginning of the normal wet season to reduce the need for irrigation.

Last Thoughts

Growing lawn in good, nutrient-rich soil that is neither too compact nor too loose, and neither too sandy and loose nor too clayey and dense, allows grass to develop deep, healthy roots. This facilitates water percolation in times of excessive rain or flooding, opens pores for optimal microbial activity and oxygen movement, helps the grass withstand drought, and enables maximum carbon storage.

ACTION TOPIC I-6 **Replace Some Lawn with Meadow** »»

Why This Matters

In nature, meadows are perpetual grassland ecosystems maintained by environmental factors that restrict the growth of woody plants. For instance, coastal meadows are constrained by salt sprays and/or saline water; arid meadows are held in check by low precipitation; and in prairies, woody plants are limited by periods of severe drought, grazing, and (historically) fire.

In our domestic and managed landscapes, small (or large) meadows can be an excellent climate-wise alternative to conventional lawn. They provide a green sward of low- or moderate-height vegetation that keeps a space open, as lawns do, but they require no mowing or standard lawn care, and they usually provide habitat for birds, pollinators, and other beneficial insects.

Urban and suburban meadows, like natural meadows, will require some action to keep the vegetation mostly herbaceous. In many cases, annual

Wildflower Meadows in Golf Courses

New golf course research at the University of Florida is making the case for installing wildflower meadows in out-of-play areas. These ecosystems not only increase the beauty of the landscape and reduce the mowing acreage, but initial results also show that beneficial insects, attracted to the wildflowers, prey on turf pests. By reducing the use of pesticides, golf courses can save money, be more attractive, and be more environmentally healthy, which is good for everyone.

mowing or hand weeding is enough to prevent woody plant growth. However, the maintenance of meadows is a large and complex subject. We offer some general guidelines below, but getting the advice of a local expert will go a long way toward ensuring success.

Meadows provide many benefits. They provide wonderful habitat for butterflies, birds, and the many insects (and caterpillars) that birds need as food. After meadows become established—assuming the proper species have been chosen to match the conditions—they will need little or no watering. Sometimes a meadow can serve as a foundation planting in place of conventional shrubs that might outgrow the space and need regular trimming or periodic replacement. And perhaps one of their least-recognized but most important attributes is that their long-lived species, notably perennial grasses, develop extensive roots and complex soil ecosystems, both of which substantially increase carbon storage in soil.

Actions

There are several ways to create a meadow in urban/suburban landscapes. No matter which you choose, be aware that a new meadow will probably take several years to become a relatively stable ecosystem. Establishing a meadow takes a bit more care than lawn. And meadow maintenance, while much less intensive than lawn care in terms of time, energy, and fossil fuels, requires a bit more knowledge and attention, a small price for the great pleasure of having a vibrant meadow in the landscape.

Generally speaking, a mixed meadow (one with native grasses and wildflowers) usually works best when it attracts both pollinators and birds. The flowers provide valued nectar and pollen (and beauty), and the grasses hold the soil, provide winter texture, reduce places for weeds to grow, and physically support the wildflowers. In addition, the grasses, when left un-mowed over the winter, provide cold weather food and habitat for many species, including early-migrating and nesting birds.

⊙ **Choose a suitable location.** Sunny or mostly sunny areas work best for meadows. Some seed mixes, however, are specifically designed for partial shade, such as you might find near a house or at the edge of a woods; be sure to select this type of mix if you want to grow meadow plants in areas that receive less than 4–6 hours of sun a day.

If the meadow will be started from seed, installation will be easiest on level or nearly level land. On steeper slopes, some kind of erosion control material may be required to prevent the seed from washing downhill in rain.

⊙ **Choose the best time of year.** In general, the best time to transform a lawn to a meadow corresponds to the best time to plant wildflower seeds for your climate. In the southeastern states, this is in the fall; in more temperate regions, it may be at the end of winter. Of course, if the meadow will be created by simply stopping the mowing of a lawn, any time is good.

> If suburbia were landscaped with meadows, prairies, thickets, or forests, or a combination of these, then the water would sparkle, fish would be good to eat again, birds would sing and human spirits would soar.
>
> —Lorrie Otto,
> *Wild Ones Journal*, 2003

What about Tidiness?

It's possible that neighbors, accustomed to neatly trimmed landscaping, might feel unsettled by a meadow that appears untidy or unkempt. Possible solutions, also called "cues to care" (see IV-3 on page 113 for more on this subject) include:

- *Create a "civilized-looking" edge* in the most visible areas by planting a border of showy bunching grasses and/or low-growing shrubs. Use plants that are large enough to be immediately recognized as a border—or just regularly mow a narrow strip all around the edge.
- *Create paths through the meadow.* These could just be mowed routes that might change each year, or they could be more permanent, with stepping stones in a grass, deep mulch, or a pea stone footpath.
- *Add familiar garden features* such as birdbaths and birdhouses, beehives, fences, garden ornaments, a gazebo, or perhaps a bench under the shade of a tree or pergola.
- *Put up a sign* that explains what's going on: "Pollinator Meadow," "Wildlife Habitat Landscape," "Bird-Supporting Garden," or other similar messages.

> By planting a meadow or prairie, you become a member of a landscaping movement that has endured for over a century, one that embraces diversity and the natural world. You join like-minded caregivers who believe we must take immediate action to restore habitat, and become stewards, not dominators, of ecosystems.
>
> —Catherine Zimmerman,
> *Urban and Suburban Meadows*, 2010

Starting a meadow from an existing lawn is the most climate-wise method, and it can be done in one of two ways:

- **Just stop mowing.** This strategy can work for a lawn in any state of health—thick and lush, or thin and straggly. At first, the growth will be fairly uniform, but within a few months different types of plants will show themselves, especially if sustainable lawn care without pesticides has been practiced for a year or more. While a lawn-started meadow will look weedy at first, it is definitely the easiest way to start.

- **Scalp the lawn.** If the lawn is thin, the other option is to mow the lawn to 1 inch or shorter. Rake up all the clippings and loosen the soil as you rake. Time this mowing so that it is the ideal time to sow native wildflower seed for the region. In the South, fall is the best time, since the grass will be going into winter dormancy. Irrigate daily for a week, and then gradually cut back to little or no irrigation.

With either method, since the existing grass is not killed, these meadows will have a relatively high grass population, but some grass could be replaced over time in selected areas with forbs (non-grass flowering plants). Plants

Credit: SReed

Figure I-16: The grass in this large side yard was left to grow long and is now mowed just once a year. Beehives and an attractive movable stairway signal to any concerned neighbors that this is not just abandoned lawn.

or plugs will work better than seed in this scenario.

   Note: If the site contains an extensive population of invasive plants, this method is not appropriate, but if there are only a few areas of undesirable or aggressive plants, they can be removed by hand at the start.

> **Start a meadow by eliminating all existing vegetation.** There are several ways to do this, with pros and cons for each method. Whichever method you chose, the idea is to create a more or less clean slate, to give your seeds or plants a better chance of survival. (See the sidebar on page 26 to learn about methods for removing lawn.) If your site presently contains many annual weeds or invasive plants, consult with a local agricultural extension or natural resources agent to determine best strategies for removal.

> **Plant the meadow.** A wildflower meadow is a complex ecosystem consisting of annuals, biennials, and perennials and will vary greatly in different climates. In a natural environment, a meadow may consist mostly of grasses with only a 20% or less coverage by forbs (flowers). In a created

Figure I-17: Here, in a damp swale that was often difficult to mow, the lawn was allowed to develop into a wet meadow. Native blue vervain and goldenrod share the space with Eurasian tansy (the yellow dots), which is somewhat invasive and might in time, without human intervention, outcompete the native plants. Time will tell!

Figure I-18: This New England meadow emerged from what used to be lawn, with the addition of only a few plants such as this Meadowsweet (in the foregound) and Steeplebush spirea.

Credit: SReed

meadow, many people include a higher ratio of forbs for the beauty of their flowers and habitat values—40% is a good starting point. Be sure to include several species with variable blooming times so that something is flowering throughout the growing season to serve the pollinators.

If you don't wish to decide on your own mix of plants or seeds, many companies offer pre-made meadow seed mixes suitable for various conditions; these tend to be more affordable than customized mixes, and the results can often be modified later by adding other sprigs or plants to an established meadow. In addition, the best suppliers and nurseries provide step-by-step instructions for installation, along with helpful phone advice.

Last Thoughts

To preserve the open feeling that a lawn provides, but to avoid the regular maintenance, fossil-fuel consumption, air and noise pollution, empty habitat, and CO_2 emissions that a lawn involves, consider creating a meadow. Good places for meadows could even be in our sunny front yards, many of which stand empty and unused most of the time. Enlivening this space with gracefully waving grasses, drifts of meadow flowers, and lots of pollinator visitors would be an excellent way to display our intentions to take care of the environment and be leaders in the movement to help reverse climate change.

ACTION TOPIC I-7 Ideas for Large/Public Lawns ≫

Why This Matters

Churches, schools, businesses, municipalities, states, and other larger landowners often possess huge swaths of unneeded and unused lawn. If homeowners can make a real difference by replacing most of the lawn on their own small properties, imagine the multiplied effect if land managers changed their status quo from large lawns to something else. We'd have less pollution, fewer greenhouse gas emissions, reduced stormwater overflow, and at the same time we'd have richer habitat, more food production, and a cooler environment.

Landowners also stand to benefit from the switch. In addition to being better stewards of their land, large landholders could save money that could be spent elsewhere. Also, if the community becomes involved in the project in some way, they could generate goodwill and positive publicity, which might influence even more homeowners and other businesses to follow suit.

Actions

Large landholders may wish to replace the majority of their expansive, uninteresting, and expensive-to-maintain lawns with community gardens, butterfly gardens, vegetable gardens, fruit orchards, rain gardens, or wildflower meadows.

> **Use new plantings to cool buildings and reduce energy consumption.** Strategic plantings of trees and shrubs on the southern and western exposures of buildings can reduce the need for air conditioning. Plantings around air compressors produce shade that can help the machines to operate more efficiently. (See VI-4 on page 168 for more ideas.)

> **Replace some lawn with rain gardens.** Businesses and commercial parks looking for ways to be more efficient and to contribute to the community could use

Credit: GStibolt

Figure I-19: Florida has a county-by-county roadside meadow program. In this case, the adjacent farmer has also planted wildflowers, which made quite a show.

more sustainable landscape practices around their buildings. Installing rain gardens to capture their stormwater runoff before it goes into retention ponds will improve the quality of our waterways.

- ⦿ **Churches, schools, and youth organizations could grow edibles.** Think of all the good a church and its members could do by creating a community vegetable garden and/or a fruit orchard to replace some of their lawn. They could raise vegetables for their own families and share their excess harvest with the less fortunate. Many people who live in condominiums or in houses on small lots might love to have this chance to show their children where food really comes from. It's all part of being good stewards of the earth.

- ⦿ **Replace roadside lawns with meadows.** Many municipalities and states are already working to replace mown roadsides and median strips with wildflower meadows, groups of trees away from the road, and other low-care landscapes.

Figure I-20: The owners of this Cornwall, Vermont, bed and breakfast have stopped mowing more than half of their large lawn, and now plan to manage the land using permaculture principles.

Credit: SReed

Some training will be needed to help workers manage these areas more sustainably as the need for mowing is reduced.

▶ **Include rain gardens in stormwater management.** Climate change is likely to produce more intense storms with heavier rainfall. Instead of expanding stormwater systems, municipalities could remove lawns from median strips and replace them with rain gardens and bioswales by cutting the curbs to remove water from streets and, depending upon the location and need for visibility, they could plant trees, shrubs, and tough herbaceous plants to soak up the stormwater. As a bonus, there will be more trees to cool the air. (See more about trees in urban areas in Section VIII and more about rain gardens in Section III.)

▶ **Save taxpayer money and set a good example.** Local governments could set a good example for their citizens by replacing lawns around municipal buildings with other alternatives that cost less to maintain, improve the air and water quality, and help diversify the landscape.

▶ **Local governments could educate citizens to be greener.** Local governments could encourage gated communities within their borders to be better citizens and revise standard requirements for flawless lawns. When new developments are suggested, local governments are in a position to encourage sustainable land use patterns, including greenways, bike paths, and less lawn. They could change outdated zoning laws to allow citizens to replace their lawns (even front lawns!) with vegetable gardens. And they could set up workshops to educate their citizens on how to remove lawns as well.

Last Thoughts

If owners and managers of large tracts of lawn were to change some of their practices, this would save money, be better for the environment, and potentially contribute substantially—in proportion to the extent of lawn being managed differently—to reducing climate change. In addition, their actions could have a great impact on the larger community; users and viewers of the new landscapes could learn from these examples and be inspired to follow suit.

TREES AND SHRUBS

Overview

Trees and shrubs are among our landscapes' most valuable assets. They often represent a large financial investment on our part, and they definitely represent a major investment on nature's part, in terms of time, energy, and the nutrients expended to create them. Beyond the fact that they delight us aesthetically, trees and shrubs play many other roles in our lives, as they:

- Cool buildings, helping us save energy.
- Form the structure and framework of our outdoor spaces.
- Screen undesirable views, create privacy, and frame favored views.
- Increase property values.
- Heal the mind and spirit.
- Mark the seasons.
- Provide a sense of place, belonging, and familiarity.
- Help slow or reverse climate change by removing CO_2 from the atmosphere and storing it in their trunks, branches, roots, and the soil where they grow.

Forests constitute one of the two primary types of natural vegetation (along with grasslands) in the terrestrial areas of the natural world. In addition to— and perhaps more important than—what they give to us humans, trees and shrubs also provide these vital services to the natural world:

- Provide diverse and essential habitat: trees offer nesting places, cover, and abundant insect food; shrubs offer nesting places, pollen, nectar, fruit, and cover.

Credit: SReed

Figure II-1: Wide-spreading oak branches support abundant life and cast a generous shade without endangering the house.

- Produce oxygen, as a by-product of photosynthesis.
- Cool the air and the ground.
- Absorb polluting gases and trap microscopic polluting particles in their foliage.
- Pull moisture from the soil and release it into the atmosphere.
- Break the impact of rainfall, which can help reduce soil erosion.
- Hold soil on hillsides and slopes, and help absorb rainwater like a sponge.

What about Shrubs?

In this section of the book, much of the information might seem to relate more directly to trees than to shrubs. This is not to slight shrubs, but rather it reflects society's greater scientific knowledge about trees, the larger and more obvious impact of trees in our landscapes, and their greater capacity for carbon sequestration and effect on climate issues. However, it's important to note that although shrubs might seem to hold less carbon in their biomass than do trees, and for a shorter time, shrubs actually do contribute greatly to the amount of carbon held in soil ecosystems (for more information, see "A Primer on Soil Carbon and Ecosystems" on page 132). In addition, shrubs play a significant role in the work of nature, providing a huge

A tree's beauty is functional, like the beauty of a great machine that performs swiftly and quietly… every accessory fits perfectly into the situation which demands exactly that thing. We can justly admire man-made structures like a machine shop or milk pasteurizing plant or a rolling mill. These are, after all, only adaptations of fundamental laws. But we can be astonished at a tree which is an original expression of those laws, mysteriously compounded out of the elements.

— Rutherford Platt,
This Green World, 1942

A Closer Look at Trees

How many of these facts surprise you?

- A tree can absorb close to 50 pounds of carbon dioxide per year, so by the time it's 40 years old, a tree can sequester 1 ton of carbon dioxide.
- In a single day, one large tree can lift up to 100 gallons of water out of the ground and discharge it into the air.
- One large tree can provide a day's supply of oxygen for up to four people.
- Almost 98% (by weight) of a tree is made up of six elements: carbon, hydrogen, oxygen, nitrogen, phosphorus, and sulfur.
- Trees do not grow beyond their ability to support themselves. During periods of stress, they shed leaves, flowers, fruit, and/or branches.
- Different parts of a tree grow at different times of the year. A typical pattern is for most of the foliage growth to occur in the spring, followed by trunk growth in the summer and root growth in the fall and winter.[1]

Credit: SReed

Figure II-2: Given a chance, a whole tiny ecosystem might thrive in the protection of one mature maple tree.

> It is increasingly clear that much of our wildlife will not be able to survive unless food, shelter, and nest sites can be found in suburban habitats. And because it is we who decide which plants will grow in our gardens, the responsibility for our nation's biodiversity lies largely with us. Which animals will make it and which will not? We help make this decision every time we plant or remove something from our yards.
>
> —Doug Tallamy,
> *Bringing Nature Home*, 2007

variety of essential pollens, nectars, seeds, fruit, and nest sites, along with needed protection from predators.

In This Section

For all of these reasons, trees and shrubs are vitally important in both our individual daily lives and the larger realm of civilization and humanity. This section of the book presents ideas for helping us get the most from—and give the most to—the woody plants in our care.

Action Topics

- II-1: Take Good Care of Woody Plants
- II-2: Choose Species Able to Tolerate Changing Conditions
- II-3: Maximize Carbon Storage in Woody Plants
- II-4: Get the Most Cooling Benefit from Trees

ACTION TOPIC **II-1** **Take Good Care of Woody Plants** »

Why This Matters

Trees provide vital carbon storage and wildlife habitat for the natural world, along with all the delight and pleasure they give to human beings, so it just makes sense for us to do everything possible to protect and preserve them. And when it comes to taking care of trees, there are two things we need to pay attention to: the aboveground parts we can see, and the belowground parts we generally can't see and don't notice.

We often think of trees as mighty and strong. And it's true: trees can grow amazingly large. They can hold their great weight upright against relentless gravity, and withstand the forces of wind, deluge, blizzard, heat, and cold. But still, tree trunks and branches are not indestructible.

We also tend to imagine that tree *roots* are similarly strong, with their ability to support a massive structure overhead, and their apparent power to crack sidewalks and penetrate foundations. But although the large roots of trees are pretty tough, a tree's *feeder roots*, which actually keep the tree alive, are quite sensitive. Taking care of a tree's roots is a little harder than protecting the tree itself, because:

Credit: SReed

- The roots are invisible to us, so we tend to forget about them.
- We can't see or know exactly how our actions are affecting them.
- The care they need might be ecological and/or chemical, and this is tricky to understand and get right.
- When roots are suffering, sometimes the only evidence might appear in the uppermost branch tips, which we hardly notice until the foliage of whole branches starts to look discolored, droopy, tattered, or stressed. And by then, the damage is often irreversible.

Figure II-3: Mighty and strong they may be, but trees and their roots still need to be treated with care.

This *Action Topic* covers both the aboveground and belowground aspects of tree care. Keep in mind that when the root zone of trees is protected and left undisturbed, this also minimizes the release of soil carbon back into the air.

Actions

The actions detailed in this section fall into three categories: before construction, during construction, and after construction.

Before Construction

- Consult with a certified arborist whenever a project involves cutting into any tree's root zone.
- Never tear or rip roots with a backhoe; instead, cut all roots cleanly, avoiding any roots over 4 inches (10 cm) in diameter, as these roots are likely to

be essential support roots. Don't cut any support roots on the side opposite a nearby building because that could cause the tree to fall toward that
building.

- Maintain and/or restore the essential soil fungi on which trees' roots depend for optimal uptake of nutrients. Fungi grow naturally in most healthy,
moist, and/or mulched soil. Mycorrhizal amendments can also be made to
the soil.
- Avoid cutting roots that would make trees vulnerable to falling during
strong storms. This is especially important in coastal regions, where hurri-
canes may become stronger and more frequent.
- Add a temporary construction fence around the outer root zone area of a
single or stand of trees. The most critical area to protect is a zone that extends from the trunk outward a distance that is either one foot (0.3 meters)

Interesting Facts about Tree Roots

- The spread of a tree's branches (when grown in the open) generally ranges from 65–100% of the tree's height. In contrast, the root system often forms a root zone with a diameter of one, two, or three times the height of the tree, and sometimes much more. In a healthy forest, some tree roots will reach hundreds of feet beyond the trunk; and this is true to a lesser extent even in lawn.
- A tree's roots rarely are evenly distributed; there might be thousands of medium-sized roots and root hairs connected to the tree via a single large transport root.
- Tree roots don't grow very deep, and the majority of roots grow in a wide, flat "pancake" of soil. Most tree roots are in the top 12 inches of soil, although in rich, deep loam, roots may extend deeper.
- Roots store more starch (carbohydrates and sugars) than the trunk.
- About 60% of a tree's biomass is in the trunk. The tender feeder roots of a tree make up about 5% of the plant's biomass, and larger transport roots comprise another 15%. This corresponds almost exactly to the aboveground proportion of 5% foliage and 15% branches and twigs. In other words, only 10% of a tree is used to produce/gather the nutrients needed for growth. And half of this vital function is performed below ground! This is why it's so important to protect tree roots.[2]

of distance for every one inch (2.5 cm) of trunk diameter or to the drip-line of the canopy, whichever is larger.

- If they are small enough to be transplanted, relocate trees that are going to be fatally damaged by construction.

- Don't save isolated, spindly trees, which will be vulnerable to wind damage, disease, or infestation. If the tree is worth saving because it's in an ideal spot or if it's an unusual species with good habitat values, create a grove of other trees and shrubs around it after construction.

- Perform basic tree maintenance for all trees near the building, such as mulching (see V-5 for more about mulching), watering, and careful pruning (no topping or lollipopping).

- Consider assigning a financial penalty for harming any valuable tree on a job site. Identify these trees and bring them to the contractor's attention.

- When feasible, save a dead or failing tree as a standing snag, for wildlife habitat.

- In situations where vehicles must drive over tree roots, lay over the root zone a thick layer of spongy mulch covered with either plywood or some other traction material, to spread out the weight of vehicles (or stored material).

- In vulnerable or constrained sites, require that the storage area be as small as possible, to minimize soil compaction in nearby root zones.

Credit: SReed

Figure II-4: When moving a tree to protect it from construction damage, dig straight down to cut the roots cleanly, make the root ball as large as possible, and replant immediately.

Credit: SReed

Figure II-5: *Never build a raised bed of soil around a tree.* This conventional landscaping action will rot the tree's bark—inviting disease and pests—and deprive the roots of needed oxygen.

During Construction

Although different tree species have differing tolerances for harm to their roots, it's best to follow these guidelines. Within the canopy area of any tree:

- Don't pile construction materials, soil, or mulch, and don't park or drive any vehicles.
- For underground utilities or irrigation avoid using trenches, which neces-

---

Changing the Grade Around Trees

Adding or removing soil near a tree may seriously disturb the roots and their ability to obtain nutrients, moisture, and oxygen from the soil. If no alternatives exist, get the advice of a certified arborist. And at the very least follow these guidelines:

Raising the Existing Grade

- Never place any fill or organic materials directly against a trunk.
- Limit fill to a maximum of 2–4 inches deep, and use fill with a coarser texture than the native soil to maximize movement of air and water to the roots.
- Don't spread fill or work the ground with equipment that compacts existing soil.

Lowering the Existing Grade

Lowering the grade can be as harmful as raising it if the work involves cutting any large tree roots or more than about 25% of the feeder root area.

If the grade must be lowered, though, several months before construction, thoroughly irrigate the tree to build up root growth, then cleanly cut roots in the area where the grade is to be lowered. This work is best done by a professional. If possible, use retaining walls beyond the canopy spread to create the maximum area of unchanged grade around a tree. Mulch the exposed root area to prevent soil erosion and moisture loss and to keep temperatures low, but don't let mulch touch the trunk.

---

sarily sever all roots. Tunnels are preferable, at least 18 inches below the soil surface to protect most roots.

- Don't wash equipment, rinse concrete chutes, etc. within a tree's dripline.
- Don't let vehicles and heavy equipment break or tear branches overhead.

After Construction

Evaluate and restore trees to their best possible condition. This is especially important for the trees that will be providing valuable shade to the house (See II-4 on page 67 for more about maximizing the cooling effects of trees), or those that form an integral part of wildlife habitat or wildlife corridors in your region, or those you just love! In every case, and especially if a tree is standing within 20 feet of a house, the ideal approach is to get the advice of a certified arborist.

Last Thoughts

Trees and shrubs play valuable roles on the planet. Both provide essential habitat, both build soil ecosystems, and together they create the rich mosaic of vegetation that constitutes the working of nature across much of the earth. In our residential and managed landscapes as well, trees and shrubs can have great significance. Trees might seem more important because they represent a large investment and take so long to mature, their trunks and branches sequester tons of carbon, and their overhead canopy adds so much to our lives. But we should remember that shrubs are equally vital: their cover, nesting sites, flowers, and fruit invite countless beautiful birds and pollinators to live close to where we can see and appreciate them.

> We should forever bear in mind that the beautiful world our species inherited took the biosphere 3.8 billion years to build. The intricacy of its species we know only in part, and the way they work together to create a sustainable balance we have only recently begun to grasp. Like it or not, and prepared or not, we are the mind and stewards of the living world.
>
> —E. O. Wilson, *Half-Earth*, 2016

ACTION TOPIC (II-2) Choose Species Able to Tolerate Changing Conditions »

Why This Matters

Choosing plants that are most able to tolerate changing and extreme conditions will help ensure that our landscapes thrive in the years ahead. This is particularly important for trees and shrubs because they are such a long-term investment and provide such vital services. Furthermore, by taking steps to ensure that our trees and shrubs do live a long time, this will help other species in finding the ecosystems and habitat they need.

Plant Hardiness and Heat Zone Maps

The USDA Plant Hardiness Zone map tracks the minimum average temperatures in each region to help gardeners select plants or seeds appropriate for their regions. Zones are based on an *average of minimum annual temperatures* over several past years, NOT the lowest temperature ever recorded or ever expected in that region. In addition, many other factors contribute to plant survival beyond just the coldest temperature in winter. Wind, soil type and moisture, humidity, storms, duration of cold periods, and how plants break dormancy, will also affect success. Started in the 1930s, the map has been updated several times as the average low temperatures have shifted northward.

In 1997, the American Horticultural Society created a *heat zone* map that measured the average number of days over 86 degrees Fahrenheit, and they worked with growers to classify thousands of plants by their heat zones. The map is intended to show plants' heat tolerance, but it assumes that plants are receiving adequate irrigation. So, it neglects the subject of drought tolerance, which can be an important factor as we try to reduce irrigation overall, and also for choosing plants for wilder areas that will not receive regular supplemental irrigation.

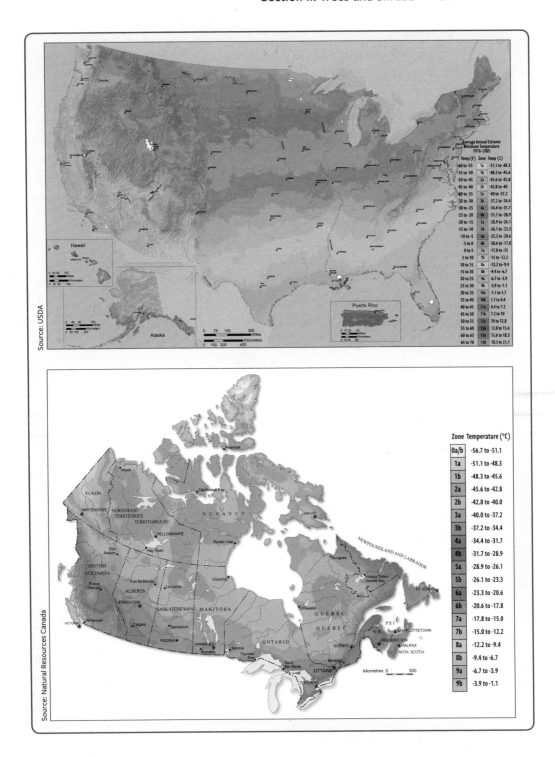

Source: USDA

Source: Natural Resources Canada

Zone	Temperature (°C)
0a/b	-56.7 to -51.1
1a	-51.1 to -48.3
1b	-48.3 to -45.6
2a	-45.6 to -42.8
2b	-42.8 to -40.0
3a	-40.0 to -37.2
3b	-37.2 to -34.4
4a	-34.4 to -31.7
4b	-31.7 to -28.9
5a	-28.9 to -26.1
5b	-26.1 to -23.3
6a	-23.3 to -20.6
6b	-20.6 to -17.8
7a	-17.8 to -15.0
7b	-15.0 to -12.2
8a	-12.2 to -9.4
8b	-9.4 to -6.7
9a	-6.7 to -3.9
9b	-3.9 to -1.1

Actions

To create optimal conditions for your landscape, take some or all of the following actions:

⊙ **Use mostly species native to your own region.** If possible, native plants should make up about 75% of a property, excluding the food-growing portions (See IV-1 on page 105 for a definition of native, and Section IX for more on growing food). Not only are these species most likely to adapt to further changes, but they also provide the best habitat for other organisms that are also trying to adjust to new conditions caused by climate change. Further:

- Choose plant species able to thrive in a wide range of soil moisture, drainage, texture, and acidity conditions.
- Choose species tolerant of urban stresses such as air pollution, night-lighting, drought/heat, and soil compaction.
- Minimize or avoid using plants that have a narrow native range, because new conditions are likely to reduce their survival.

Figure II-6: Many different hawthorns (like this one) can tolerate a wide range of soil and light conditions. Choose native species, as opposed to hybrids and cultivars, to help support the most wildlife.

Credit: SReed

- Some plants with very specific requirements might still deserve a place in our gardens, perhaps simply because they have aesthetic or sentimental value, but be aware that they may demand extra care as conditions change.
- Choose plants in the center of their native ranges, i.e., avoid plants for a location at the edge of a species' native range. For North America, species' range information is available at the USDA's plant database website: plants.usda.gov.
- Be cautious about planting species that are native to regions further south. In particular, avoid using species that are known as "spreaders" or those that tend to grow aggressively. These have greater potential to become invasive in their new location. (See the sidebar "What about Assisted Migration" on page 242.)

How Does Adaptation Work?

Adaptation is a tricky word, with several closely related meanings. When a species is *adapted* to a particular habitat or region, this means it is able to thrive and reproduce within those conditions. When we say a plant species *adapts* to climate change, this does not mean that a plant (or group of plants) changes in some way that makes it more tolerant of new conditions. Rather, it is the offspring of parent plants that make the change.

This happens because cross-pollination produces genetic variability in seeds, some of which possess characteristics that make them more or less able to thrive in hotter, colder, wetter, drier, or more or less acidic situations. Those individuals that can tolerate the new conditions and reproduce to pass on their genetic information to another generation, are *adapted* to that place. Bear in mind that it is individuals, not a species as a whole, that adapt. And such adaptations—along with natural selection and evolution—result only from cross-pollination and the presence of genetic diversity within populations (see IV-4 on page 117 and IV-5 on page 123 for more on this subject).

❯ **When buying plants, select individuals with healthy root systems.** This will maximize their ability to absorb nutrients and water, which is so important when times are tough. Buy plants that have been bred in a nursery, not dug from the wild.

Credit: SReed

Figure II-7: All across the North American West and Southwest, poplar trees rank among the most tolerant of drought and erratic moisture conditions.

In regions experiencing high heat and drought, look for plants with drought-tolerant characteristics, including:

- Small leaves
- Waxy leaves
- Fleshy leaves
- Leaf hairs, because they reflect light and catch moisture
- Deep leaf lobes
- Plants that typically grow on slopes or cliffs, where moisture is often scarce
- Plants typically found in sandy soil
- Trees/shrubs with upright crowns instead of spreading crowns

For more information about landscaping for drought, see III-1 on page 77.

In regions experiencing more intense rain, flood tolerance is the key characteristic to look for—and it is even more complex than drought tolerance. The limiting factor in a plant's tolerance for flooding is not necessarily simply the presence of excess moisture; the problem is usually the plant's lack of access to oxygen in the soil. Oxygen must be absorbed through root hairs to power metabolic processes, which enable the plant to take up nutrients from the soil. In fact, flood tolerance is essentially a species' ability to cope with low soil oxygen.

Figure II-8: Like many plants that thrive in wet soil, the native highbush blueberry is also quite capable of growing in average or even dry soil.

Trees that tend to be capable of extreme flood tolerance include typical wetland species such as larch, tupelo, willow, and bald cypress. Those that can tolerate moderate flooding include red maple, boxelder, silver maple, alder, river birch, honey locust, sycamore, sweetgum, and all hawthorns. Bear in mind that before any of these species are purchased and planted, they should be further evaluated for their suitability to any given situation by checking with local native plant societies or garden centers.

For regions experiencing more drought *and* flooding, some of the wetland species just mentioned can be a good choice. Many flood-tolerant species can thrive in average or dry soil; they can do this because their roots are able to use anaerobic respiration to obtain the needed nutrients in saturated, oxygen-depleted soils. Dry soils are not oxygen-depleted, but a shortage of moisture does impair plant roots' ability to take up nutrients. As a result, floodplain species can sometimes be a good choice for dry and average landscapes that occasionally get flooded.

In contrast, trees and shrubs adapted to dry soil conditions will generally not thrive in floodplains, swamps, or wetlands, or during prolonged flooding during the growing season. (For more about flood-proofing a landscape, see VI-5 on page 171.)

Last Thoughts

In the coming years, it may be difficult to alter our aesthetic preferences and desires for our landscapes. But as conditions become more unpredictable and challenging, this kind of adjustment is likely to be necessary in some regions—if we want our landscapes to thrive and give us the pleasure we hope for from them.

> The garden of the future will be a shade garden. There are many reasons: fiscal, historical, environmental—and for the sake of our health and of the planet's. In many ways, this is an older notion of landscaping—it's planting for generations to come. It won't happen overnight, though. For one thing, it means planting trees.
>
> —Ken Druse,
> *The New Shade Garden*, 2015

ACTION TOPIC   **II-3**   **Maximize Carbon Storage in Woody Plants**   ≫

Why This Matters

To curb global warming, humanity's greenhouse gas emissions will have to slow or stop altogether. Many governments, organizations, and businesses are working toward this goal. In addition, there are many steps we can take as individuals to emit less CO_2 into the atmosphere.

To actually succeed in reversing global warming, though, we also need to find ways to *remove* excess CO_2 from the atmosphere and *sequester it* in some form of long-term storage. Again, many large-scale solutions have been proposed for sequestering carbon, and some of them will undoubtedly prove effective in time. But even in our own private and public landscapes, we can help take carbon out of the atmosphere and stash it away where it won't do harm. How? One way is by growing more woody plants. (Another solution involves storing more carbon in soils; see Section V.)

How do trees and shrubs store carbon? Like all green plants, trees take in carbon dioxide during photosynthesis and use it to make sugars/starches, which they then use for energy and to build their wood and roots (biomass). These plants continue to add carbon for as long as they are growing. And they lock this carbon in their wood for as long as they are intact. Although they do release some carbon dioxide as a result of daily respiration and eventual decay, healthy, mature trees in forests typically store carbon at a greater rate than they release it.

Note: Carbon is held in wood even after a plant dies. So by using wood—ideally locally harvested or recycled wood—for construction, we continue to sequester that carbon.

Credit: SReed

Actions

In addition to large-scale actions such as reducing worldwide deforestation and adding forests wherever possible (especially in cities), there are many things we can do to help store more carbon in trees. Here are a few of them:

❯ **Take care of trees** so they live as long and healthy a life as possible. (See II-1 on page 52 for specific suggestions about how best to care for trees.) Remember that old trees hold carbon from decades past, and keeping

Figure II-9: Apparently immune to Dutch Elm Disease, this magnificent elm tree in Amherst, Massachusetts, reminds us that all species in nature contain individual organisms, each with its own unique characteristics.

How Much Carbon Can Woody Plants Sequester?

The answer depends on many factors, most of which are fairly difficult to quantify, including:

- Diameter, height, age, and rate of growth.
- The weight and density of the wood itself (which depends on species and rate of growth).
- Conditions in which the plant is growing (temperature, soil type, season, competition, available space).
- Relative size of root mass compared to aboveground mass.
- The formula being used for the calculation.

Here, we do not here take a position on exactly how much carbon any given tree or acre of trees will sequester. Many organizations have attempted to distill the available data into a simple set of guidelines, but the results vary quite widely. Online "carbon-calculator" websites can, however, provide a *general* idea of the amount.

that carbon safely stored out of the atmosphere is one of our best tools for reversing climate change.

🔾 **Plant as many trees as possible.** However, make sure to put them in places where their mature size won't be a problem (that would cause them to be cut down).

🔾 **Plant a diversity of trees** well-suited to the region and soil type, so that if one species suffers from a pest infestation or disease, the remainder of the trees will continue to grow and store carbon.

🔾 **Minimize damage to trees** before and during new construction by accommodating the trees' sizes in the plans and by fencing them off to reduce root zone damage.

Credit: SReed

Figure II-10: All beech trees, including this European "Copper" beech, are vulnerable to a serious beech bark disease, which may be worsened by warmer winters. Arborist treatment can help save valuable landscape specimens.

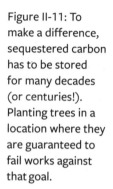

Figure II-11: To make a difference, sequestered carbon has to be stored for many decades (or centuries!). Planting trees in a location where they are guaranteed to fail works against that goal.

Credit: SReed

> **Grow a mix of fast and slow-growing trees.** This is important because while fast-growing trees tend to store carbon sooner, slower-growing trees species tend to be the ones that live longer, providing their valuable shade, habitat, and carbon sequestration for the long run.

> **Create natural woodland *mini-ecosystems*** that build carbon in the soil. Trees planted in groups or groves, together with shrubs, will provide better soil conditions for all, along with better habitat for wildlife, and more wind tolerance.

Innovation: Enhanced Carbon Storage

Those who own large acreages of harvestable timber might explore *enhanced carbon storage* forest management practices. These techniques aim to maximize both the volume of harvestable biomass and the quantity of carbon in the soil. The methods involve "nudging" forests toward old-growth characteristics such as structural complexity, mixed-age stands, and a wide diversity of habitats. This new idea holds potential for landowners who wish to take care of the planet but also need to gain a financial return on their woodlands.

Groves also require less ongoing maintenance (see more about this in IV-2 on page 109).

▶ **Use wood products in structures** designed to last many years. This prevents the breakdown and decay that release CO_2 back into the atmosphere.

▶ **Manage woodlands and forests through sustainable practices** that keep trees healthy and promote vigorous growth. Healthy trees are also more resistant to pests and diseases, which may make them better able to adjust (via their seedlings) to the stresses of climate change.

Last Thoughts

There's abundant scientific consensus that planting trees is one of the most effective—and cost-effective—ways to sequester CO_2 from the atmosphere. But remember: this solution depends on making sure the stored carbon stays out of the atmosphere. We can do this by planting long-lived trees, taking best care of them, and planting them in places and conditions where they (and their roots and the soil in which they're growing) can remain undisturbed for decades, or even centuries.

> For each acre of forest protected, the threat of deforestation and degradation is removed. By protecting an additional 687 million acres of forest, this solution could avoid carbon dioxide emissions totaling 6.2 gigatons by 2050. Perhaps more importantly, this solution could bring the total protected forest area to almost 2.3 billion acres, securing an estimated protected stock of 245 gigatons of carbon, roughly equivalent to over 895 gigatons of carbon dioxide if released into the atmosphere.
>
> —Paul Hawken, *Drawdown*, 2017

ACTION TOPIC **II-4** Get the Most Cooling Benefit from Trees ≫

Why This Matters

According to the EPA, shade trees lower the amount of solar radiation that reaches a building's walls and roof, potentially reducing indoor temperatures by 20–45°F. In addition, the roots of trees create pores in the soil that let water seep in, which helps cool the ground around a building. And trees' foliage cools the air via transpiration. In fact, trees are one of the most cost-effective,

Credit: SReed

environmentally friendly, and enjoyable ways to reduce the amount of electricity needed for air conditioning.

Note: We do recognize that, especially in the warmest regions, trees can't wholly eliminate the need for air conditioning. But they will reduce it, and if trees or shrubs can be positioned to shade the air conditioning units, their efficiency will increase.

There are various ways to maximize the cooling effect of trees. The suggestions presented here apply across much of North America; exceptions and details applicable to specific regions are noted where appropriate.

Actions

⊙ **If any large trees already shade a building, preserve and take care** of them, unless they are a species known to be invasive in your region.

⊙ **If planning for new construction** (house, shed, driveway, anything), and shade trees already exist in the landscape, design the new construction specifically to protect those trees and take advantage of their future shading benefits.

Figure II-12: Throughout the American South and Southeast, homeowners appreciate the value of towering shade trees that block the sun but allow breezes below.

⊙ **Get a basic understanding of how the sun's shifting position in the sky** affects the length and direction of shadows.

⊙ **Figure out north and south relative to your home or building.** Many websites explain how to do this. Also, notice which rooms tend to heat up the most, to determine where you need shade trees the most.

For the south side of a building:
 • Plant small trees or large shrubs (20–25 feet tall) about 5–10 feet away from

Trees and Cooling

- Air temperatures directly under trees can be 25 degrees cooler than air temperature above nearby blacktop.[3]
- There are about 60–200 million spaces along our city streets where trees could be planted. This translates to the potential to absorb 33 million more tons of CO_2 every year, and save $4 billion in energy costs.[4]
- A tree with a 30-foot crown transpires about 40 gallons of water a day, poten-

tially reducing nearby air temperatures by 2–9°F. This also adds valuable moisture to the global rain cloud system.[5]
- If you plant a tree today on the west side of your home, in five years your energy bills should be 3% less. In 15 years, the savings will be nearly 12%.[6]
- The net cooling effect of a young, healthy tree is equivalent to ten room-size air conditioners operating 20 hours a day.[7]

a building, especially in front of windows, so their overhead foliage will shade walls and windows from the nearly vertical rays of midday sun.

- Use caution when using large trees for shade on the south, because in order for their shade to have a cooling effect, the trees might be so close that large branches could endanger the roof, and tree roots might interfere with the foundation.
- In cooler regions, place evergreens no closer than 3–4 times their maximum height away from the building; this prevents them from blocking valuable mid-day warmth in winter. In warmer climates, those southside evergreens would be valuable additions to help cool the house.

Regional note: In humid areas, keep trees/shrubs far enough away from building walls to allow for

Credit: SReed

Figure II-13: A small, delicate deciduous tree (like this serviceberry) planted near a building's south wall can block the direct overhead sunlight while still allowing views and breezes below. In winter, thin bare branches barely obstruct the welcome low-angle sun of midday.

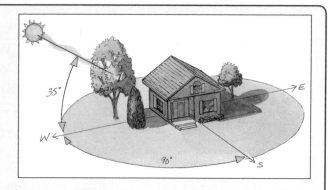

Figure II-14a: This diagram depicts the sun's position four hours *after* noon, during the summer solstice at a geographic location of 40° north latitude. The same angles would occur four hours *before* noon. At these times of day, shadows cast by trees are relatively long.

Figure II-14b: Two hours before and after noon, in the same location as Figure II-14a, the sun's steeper angle produces shorter shadows.

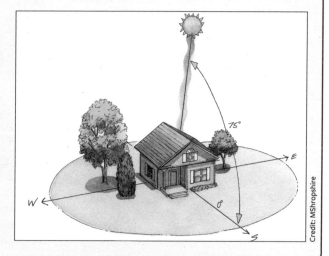

Figure II-14c: At noon, the sun is the highest and hottest it will be all day (except during Daylight Savings Time, when the sun reaches its zenith at 1 PM), and midday shadows are correspondingly short.

Credit: MShropshire

air circulation, and avoid planting anything near the house that need a lot of watering.

Southwest of a building:
- To block hot afternoon sun, plant large deciduous trees 20–40 feet (6–9 meters) away.

Regional note: In hot-arid regions, where shade is as important in the morning as afternoon, planting trees 20–40 feet from southeast walls will also provide valuable cooling benefits, reducing the buildup of daytime heat. Also in fire-prone regions, avoid planting trees closer than 30 feet to a house; even more than 30 feet away avoid resinous trees such as pines or firs to minimize the possibility of damage from fire.

West of a building (and to a lesser extent, east of a building):
- Plant large trees up to 1.5 times their expected maximum height away from a building.
- Nearer to the house, plant/preserve trees with lower crowns/branches, to block the lower rays of the afternoon sun.

For more detailed guidance about how to position trees for maximum cooling, see *Energy-Wise Landscape Design* by Sue Reed (New Society Publishers, 2010).

Last Thoughts

Shading buildings as described above can significantly cool indoor air temperatures, increase our comfort level, reduce utility costs and shrink our CO_2 emissions. Of course to shade a house, and to cool a city, we could just plant trees in every possible location. But another option, when our landscapes allow, is to intentionally position trees for their maximum cooling effect.

Shading and cooling from tree canopies can reduce summer temperatures from uncomfortable to pleasant ranges. What's more, the effect is additive: the more your neighbors add trees to their properties, the cooler the air mass within your neighborhood will be and the longer it will stay cool during extended heat waves.

—Rick Darke and Doug Tallamy, *The Living Landscape*, 2014

WATER

Overview

Climate change is likely to produce an odd juxtaposition of more frequent and longer droughts, and more frequent and high-intensity rainstorms, often coming at unusual times of year. For our landscapes, this means some of us will need to prepare for *shortages* of water, while others will have to plan for *excesses* of water, and still others will ultimately be dealing with *both*, in the same place but at different times of year. As a result, this Section addresses the dual subjects of water conservation and stewardship of stormwater.

Water Conservation

Usable freshwater is a limited resource, but until now we have tended to take it for granted. With intensifying droughts, however, our attitudes about the availability and cost of water will necessarily shift. We can and should save water in many ways indoors, but we can make a much bigger impact by transforming entire properties into water-wise landscapes.

Stormwater Stewardship

Specialized strategies come into play with excesses of water instead of shortages. Often called *rainscaping* or *green infrastructure*, these stewardship practices increase the ability of the ground to absorb, soak up, or otherwise delay the runoff of rainwater. Especially useful in urban and developed land, water-holding and water-absorbing features bring together natural and engineered elements to create systems that mimic natural processes.

Improving the ability of landscapes to slow and absorb rainwater brings several benefits:

- It reduces pollution of local waterways by minimizing the amount of eroded soil, organic matter, pesticides, fertilizers, and vehicle drippage carried into storm drains. Any water that does reach waterways will be cleaner, because particulates and other contaminants will have settled out.
- If enough people capture and slow the runoff from their own property, in the future, municipalities might not have to expand stormwater infrastructure to meet the increased volume due to severe flooding (and even rising sea levels) caused by climate change.
- Groundwater or aquifers may be more effectively recharged.
- Water features that support vegetation will also promote biodiversity.
- Finally, green infrastructure plantings can lower soil temperatures, which will help reduce the urban heat island effect. (The subjects of urban heat, green roofs, and pervious pavement are covered in Section VIII.)

In This Section

These *Action Topics* address water conservation:

- III-1: Make Landscapes More Drought-Tolerant
- III-2: Use Climate-Wise Irrigation Methods

These *Action Topics* help us deal with *too much* water in the landscape:

- III-3: Install Water-Collection Systems—Rain Barrels and Cisterns
- III-4: Add Water-Absorbing Features—Rain Gardens, Bioswales, and Drywells
- III-5: Create or Maintain Water Features—Ponds, Fountains, Pools, etc.

"Water and energy are resources that are reciprocally and mutually linked. Energy efficiency initiatives offer opportunities for delivering significant water savings, and likewise, water efficiency initiatives offer opportunities for delivering significant energy savings. In addition, saving water also reduces carbon emissions by saving energy otherwise generated to move and treat water." [1]

A PRIMER ON...

Water Chemistry and Plants

Life on this planet depends on water's unique properties. Made of two hydrogen atoms and one oxygen atom (H_2O), water molecules have a polarity that makes them act like little magnets: they are attracted to each other. This cohesion is what causes water to bead up into droplets.

Water's polarity also makes it a good solvent, enabling it to break apart, absorb, and carry organic materials such as sugars, other carbohydrates, and nutrients. Knowing how plants deal with and take advantage of water's unique chemistry can help us be more efficient gardeners.

Osmosis and Root Hairs

Near the tip of each root, plants have thousands of single-cell extensions on the root's surface, called *root hairs*. Their cells have a semi-permeable membrane that allows water and dissolved nutrients to flow into the cell.

When there is less water within root hairs than in the surrounding soil, water will flow across the cell membrane to equalize the pressure. Once water fills the root hair cells, it will pass into neighboring cells and build up root pressure, which pushes water up into the plant. Then the process of transpiration, as discussed below, carries the water farther up the plant. (See Section V for information on how mycorrhizal fungi help plants absorb water.)

Root hairs usually last just a few weeks before they fall off or are reabsorbed into the root tissue. A root must be growing in order to develop new root hairs. As gardeners, we need to keep this in mind: when irrigating plants, we need to supply water to the region where new roots are growing. Roots of established trees and shrubs might be growing many yards away from the trunks or stems.

When a plant is transplanted, most of its delicate root hairs are rubbed off as soil falls away from roots, so these plants need a lot of water in the planting hole and frequent irrigation until new root growth begins and the plant regains its network of root hairs. The larger the plant, the longer the time period when additional irrigation is necessary for the establishment of that plant. This is true for drought-tolerant plants as well. They only become truly drought tolerant after they are fully established.

Transpiration

After water enters root tissues, root pressure begins to push the water higher in the plant. But it would not go very far up the xylem (the water-carrying cells in plants) without the suction effect of evaporation through the pores in the leaves and stems (called *stomata*). The suction is caused by the differential of the concentration of water in the air and the plant: since there is less water in the air, nature works to equalize the concentrations by pulling the water from the plant.

More than 90% of the water that enters a plant runs straight through it and evaporates into the air. The volume of transpiration is greater for plants with larger leaf surface area. A full-grown oak tree can transpire more than 40,000 gallons per year.[2]

In addition, because transpiration is a set of chemical reactions that are powered by heat energy in the air, transpiration actually cools the surrounding air. As gardeners, we can use this information in designing our outdoor spaces and placing plants near buildings to reduce the need for air conditioning (see II-4 on page 67 and VI-4 on page 168).

The transpiration rate is also important in planting rain gardens. Rain gardens are designed to collect rainwater in swales and other specially designed depressions, where some of the water will percolate into the soil, and the remainder will be absorbed by the plants. Rain garden plants with higher transpiration volumes (those with more leaf surface area) will increase the water absorption rate.

The 10% of the water that is absorbed into the plant's cells serves several purposes: it carries nutrients, keeps the cells turgid, and enables the plant to photosynthesize. When soil is dry, this whole process slows down. The guard cells around each individual leaf pore are highly sensitive to water supply. When they become flaccid, the pores close up and evaporation is slowed, to protect against wilting. Gardeners need to pay attention to the wilting of plants, especially seedlings, and irrigate before permanent damage is done to the plant's tissue.

Note: Water flows from the roots upward through the plants during transpiration using tubular cells called *xylem*. Sap, which includes nutrients and sugars, is carried in cells called *phloem*. The phloem cells, which are right beneath the bark, can direct the sap upward or downward depending on the season and the needs of the plant. Girdling a tree destroys the phloem; a girdled trees dies within a year because the sugars cannot be transported to the roots or back up to the leaves in the spring.

ACTION TOPIC **III-1**   Make Landscapes More Drought-Tolerant »»

Why This Matters

Methods for building a more drought-tolerant landscape will differ depending on local climate. A landscape in a temperate zone that receives 30 or more inches of rain evenly dispersed through the year should look quite different from one in an area that receives less than 5 inches of rain per year. When we modify our landscapes to be more water-wise, they also become more sustainable for the climates in which we live. (See Section I for more ideas on sustainable lawn care and lawn substitutes.)

Creating more resilient landscapes that can withstand water shortages, water-use restrictions, and periods of hotter weather is one of the biggest steps we, as homeowners and property managers, can take to reduce our environmental footprint. In addition, such landscapes will save us money and time: drought-tolerant plants will last longer in the natural conditions of the local climate.

Actions

◉ **Choose drought-tolerant plants.** Wherever possible, use plants that will be able to survive droughts. Study local natural or untended areas to find which plants can survive on their own. See II-2 on page 58 for a list of plant characteristics that increase drought-tolerance.

◉ **Choose native plants.** When choosing trees, shrubs, and herbaceous plants for the landscape, native plants are a good place to start because they have evolved to thrive in local rainfall patterns—not only the total rainfall,

Figure III-1: Bluebonnets and blanket flowers are good drought-tolerant choices for this Texas landscape.

Credit: GStibolt

Xeriscaping

In the 1970s, in Denver, Colorado, which is in an arid region, someone coined the term *xeriscaping*: it is a combination of two Greek words: *xeros* (dry) and *scape* (view). Some people have also called it *zero-scaping* to indicate landscapes that need no irrigation. The term has now become a general gardening philosophy—the designing of landscapes that work well in their local climates without extra irrigation. Technically, a landscape in a non-arid region can't be a true xeriscape, because it's not dry. But all landscapes can be designed to thrive with only the natural precipitation of that region or just a slight amount of irrigation during extreme drought.

Credit: GStibolt

Figure III-2: This Florida pond-side planting replaced a lawn that extended right to the water's edge. Now moisture-loving plants provide habitat useful to aquatic species and other wildlife.

but also the seasonality of the rainfall such as wet and dry seasons. (See "A Primer on Native Plants" on page 103 in Section IV: Ecosystems.)

⊙ **Choose more long-lived plants.** Perennials and woody plants can develop extensive root systems that efficiently soak up moisture from the soil. We can help this process by encouraging roots to grow outward by adding a top dressing of locally made composts spread beyond the root ball of the plants we install. This action doesn't apply to annuals, which finish growing before their roots extend very far.

⊙ **Plant in groups.** The plants will be more drought tolerant together. As discussed in Sections II and VII, install groupings of plants that approximate natural arrangements in the local area. A group of plants such as a thicket or grove is generally more drought tolerant than single or specimen

Credit: GStibolt

Figure III-3: Lurie Garden in Chicago, designed by Piet Oudolf, looks like a prairie has been plunked into the middle of the city. It teems with pollinators and birds, along with people who can learn about the native plants from signs and guided nature walks. The garden, which appears to be much larger than its 3.5 acres, inspires many to re-imagine what a beautiful garden can be.

plants for a number of reasons—much of it has to do with multiple plants sharing soil spaces and interacting within the soil's ecosystem.

- **Install plants with similar water needs together.** Design the whole landscape to recognize the overall water and drainage patterns. If a group of plants needs more moisture, strategically position them to take advantage of runoff from roofs and driveways or overflows from rain barrels, or plant them in a swale.

- **Avoid thirsty lawns and other moisture-loving plants.** Avoid lawns entirely in arid or semi-arid climates. In areas that receive at least 20 inches of annual rainfall, keep lawns small and manage them sustainably with a diverse mixture of mowable plants. (See Section I for more on lawns.) Unless there is a water

> The point is, there's no one "right way" to plant a garden that saves water. Rather, it's about planting thoughtfully, using plants that can subsist on natural rainfall or judicious supplemental watering, grouping plants by water need, and saving higher-water plants you love for small areas where you can most easily enjoy and water them.
>
> —Pam Penick,
>   *The Water-Saving Garden*, 2016

feature in the landscape or some other reliably moist area, avoid plants that need regular supplemental irrigation.

Last Thoughts

Our perception of what makes a landscape "beautiful" has shifted in recent years from thirsty, perfectly groomed formal gardens to less formal, more naturalistic, and drought-tolerant landscapes that provide ecosystem services to local and migratory birds and butterflies, using only natural rainfall.

ACTION TOPIC **III-2** **Use Climate-Wise Irrigation Methods** »»

Why This Matters

Fresh water is a limited resource and is likely to become more so in an era of climate change. In addition, it takes a fair amount of energy to treat and deliver it for use in homes and businesses. But easy-to-accomplish landscape changes can make a big difference in our water consumption.

Actions

▶ **Arrange plants by watering need.** In her book *The Water-Saving Garden*, Pam Penick suggests a *ripple-zone* watering scheme in which plants needing more irrigation are located closest to the house; plants needing less irrigation are farther away, until the most outlying regions would need no irrigation ever.[3] Food crops will need much more water than the rest of the landscape, so design your method of irrigation to take care of crops first, if they are part of your landscape. As discussed in Section IX, every pound of food we grow prevents two pounds of greenhouse gas emissions.

▶ **Install a rain gauge.** Before doing any irrigation, find out how much rain you receive on an ongoing basis. The local weather reports provide only the average rainfall; it can vary widely among localities. Purchase a good rain gauge and mount it on a post away from buildings or trees and outside of irrigated areas. Record weekly and monthly totals. If you've received a half inch of rain, most of your landscape will probably not need any more than that for the next seven days or more.

If irrigation is needed, it's best to irrigate deeply, but less often. A good rule of thumb is to aim for a half inch of water for each session, which equals 2.4 gallons per square yard—important to know for hand watering. (Of course, the amount needed will vary somewhat with the climate.)

❯ **Choose the right irrigation methods.** In every landscape, there can be multiple options for how best to water plants and gardens. The following are listed from most to least climate-wise:

- *Hand watering with a watering can* is the thriftiest, because the gardener puts the water exactly where it is needed. Using a rain barrel makes it even more frugal. If a watering can's volume is three gallons, emptying it on a 3-foot by 4-foot bed will approximate a half-inch rainfall, a good amount for a deep watering.
- *Container gardens* are often watered by hand, but since the volume of soil is relatively small, they may need more frequent irrigation. Reduce the frequency of irrigation by using large containers or those with self-watering reservoirs that only need to be refilled once a week or so to keep the containerized plants moist, but not wet.
- *Drip irrigation systems* are generally more climate-wise than aerial sprayers because the water is directed only to the soil near those plants that require extra water, and there is very little water evaporation. They can be set up with a timer from a house faucet or from elevated rain barrels, or they could be part of a landscape-wide automated irrigation system. They are particularly useful for edible crops and container gardens.
- *Hauling hoses around the landscape* (for either hand spraying the gardens or setting up a sprinkler) is another manual irrigation method, but since the hoses are usually hooked up to the household water supply, this is rarely a best practice in a climate-wise landscape. If you do hook up a sprinkler, calculate the equivalent rainfall by laying straight-sided containers in several places and then running the sprinkler for ten minutes. Measure the depth of water in each container and calculate the average, so then you'll know long to irrigate to reach your goal. It's a good idea to hook up a timer to the faucet so it will automatically turn off the water flow.

Credit: GStibolt

Figure III-4: Micro emitters spray water at low pressure. Compared to high-pressure spray systems, this drip irrigation reduces evaporation and directs water only near the plants where it is needed. Temporary drip systems can be installed to provide the extra irrigation needed for the establishment of newly installed plants.

Credit: GStibolt

• *Automated irrigation systems with controllers* that can be set up to run on certain days of the week make it very easy to set it and forget it; although they require a minimum of effort, these systems can tend to waste water because they run on a schedule regardless of rainfall. There are ways to combat this waste, however. Many towns and cities require a rain sensor so automatic irrigation doesn't run when it's raining or when the humidity is high. *Smart controllers* interface with predicted weather and do not run if rain is likely; some systems may also include soil moisture detectors to make them even smarter.

If you have an irrigation system in place, you may find that as you transform your landscape using water-wise techniques, you'll use it less and less. Also, you may need to replace sprinkler heads in a lawn to those more appropriate for non-lawn landscapes with higher vegetation levels. Automated systems may still be useful for extended drought situations (if there are no water restrictions). For landscapes in fire-prone areas, you may

Figure III-5: Beyond the obvious irony here, what might be less obvious is that in addition to irrigating mulch, this climate-unfriendly landscape contains unsustainable red-dyed cypress (see V-5), and watering mulch only encourages weeds to grow.

want a system that can soak your whole landscape if a wildfire is heading your way to potentially reduce the damage to your house. See VI-7 on page 180 for more on fire-wise landscaping.

❯ **Choose the best source of irrigation water.** The most desirable for irrigation, from the perspective of conserving precious potable water, is always the lowest-quality water available for the job. In some localities, regulations prohibit irrigation systems from being hooked up to municipal tap water if another water source is available. As above, the following sources are listed from most climate friendly to least.

- *Greywater*: This is water collected from sinks, showers, and water-using appliances (with the exception of sewage, i.e., *blackwater*). It is diverted from the septic or sewer system and may be used for irrigation (best if organic soaps and detergents are used). Using greywater recycles used water and reduces the strain on sewer and septic systems. There are several ways to set up a greywater system, but keep this in mind: unlike rainwater, greywater should not be stored for more than a day or two, so the water should be continually siphoned off for use. Local regulations vary, so be to sure check about what is allowed.

- *Lake water*: Check your local regulations first, but if your property adjoins a lake, large pond, or retention pond, you may be able to use it as a source for a pump-based irrigation system. This works best if the water levels are relatively steady.

- *Harvested rainwater*: (See III-3 on page 84 for specific ideas.) Water collected in a cistern or series of rain barrels can fill watering cans for manual irrigation, or it can be hooked up

Credit: GStibolt

Figure III-6: These elevated rain barrels, connected via pipes attached to drain holes, fill up and empty as a unit. Having one spigot makes this an ideal setup for a drip irrigation system. *Note*: Gutters were present only on two sides of this building, so a separate gutter was installed to augment the collecting capability.

to a drip irrigation system. Be sure to include a timer so the supply is not drained in one session. Elevated barrels or a cistern will provide enough gravity pressure for a drip system. If they are set at or below ground level, a pump (which could be an old-fashioned hand-pump) will be needed to run an irrigation system. A drip system will require less pressure than a spray system.

- *Well water*: In some regions, shallow wells can be dug or drilled to become the untreated water source for a pump-driven irrigation system. Local regulations may determine whether this is allowed in your neighborhood.
- *Treated/Potable water*: Whether the source of your drinking water is municipal tap water or a deep well, this is the least climate-wise source of water to be used for irrigation. If there is a water softening system to treat the well water, use a faucet that is between the main pump and the water treatment system.

Last Thoughts

Many options and strategies exist for reducing the amount of supplemental water our landscapes need. However, in most parts of the world, some amount of irrigation will likely be needed for various purposes at various times of the year. The climate-wise suggestions presented here provide alternatives to conventional watering methods, and offer ways to reduce the use of valuable potable or treated water in our gardens.

ACTION TOPIC (**III-3**) **Install Water-Collection Systems— Rain Barrels and Cisterns**

Why This Matters

Collecting and managing water is an ancient practice that has been making a comeback recently for several reasons. Water shortages and water-use restrictions may limit available water. Water with no added chemicals is better for irrigating plants; it doesn't harm microbes in soil or compost piles. Reducing stormwater outflows also reduces pollution in nearby waterways. And finally, using less potable (treated, municipal) water shrinks our carbon footprint.

Actions

⊙ **Install rain barrels or cisterns to intercept water flowing from roofs.** In the southern regions of the US, rain barrels are effective all year long; in regions where water will freeze, empty the barrels and store them—inside, if possible, so they don't crack. Mosquitoes are rarely a problem with modern rain barrel systems because screens cover the catch baskets or other top openings.

For the following discussion, rain barrels and cisterns are treated together, but there are pros and cons for using one over the other.

Cisterns are usually 250 gallons or larger. They can be made from plastic, wood, or metal and installed above or below ground. Collecting water for a cistern normally includes some type of conveyance from all downspouts. Below-ground cisterns require a pump, and they can be used all year in most climates.

Rain barrels are smaller and less expensive. They can be hooked together in a series to increase the volume of total storage. Usually made from wood or plastic, they can be purchased with all the parts installed, or homeowners can create their own system using recycled food-grade barrels and individual parts.

Collecting rainwater starts with an impervious surface, usually a roof, that has gutters or other collecting devices that direct runoff to a down-spout. If there is no collection device, downspouts usually dump the rain-water directly into the landscape, but in some neighborhoods, downspouts are extended with underground pipes that carry the stormwater out to the street or municipal storm drain.

Rain barrels or a cistern will intercept the flow of water and make it available to be used later. In general, one inch of rain on 1,000 square feet of impervious surface produces 623 gallons of water. Many rainfall calculators are available online to help you determine what size of rain-harvesting system you need.

Credit: SReed

Figure III-7: A large cistern catches water from the rooftop of this New Mexico home and saves it in an out of the way location for gardening (or perhaps limited firefighting).

> **Maximize the capacity.** To do the most good in a water-wise landscape, install several barrels for a larger volume of rainwater—one that can last through prolonged droughts. It is common to hook rain barrels together with overflow pipes. Each of the barrels in an overflow system will have its own *hose bib* (spigot) installed near the bottom of the barrel. Elevate barrels so a bucket or watering can will fit under the hose bib (two layers of cinder blocks work well). Another configuration for a series of rain barrels is to set them on an elevated deck and hook them together through holes drilled in the bottoms, each one opening to a single pipe that leads to one hose bib that empties all the barrels together. This setup is most appropriate for drip irrigation, because it provides a greater volume of water, and the elevated barrels provide gravity for the irrigation system.

Figure III-8: Three rain barrels hooked together with overflow pipes at the top of each barrel provide a capacity of 175 gallons. The last barrel overflows into two watering cans for an extra six gallons of capacity. After that, the excess flows across the cement pad to a rain garden, which overflows into a wooded area.

> **Use harvested rainwater instead of municipal water.** Potential uses include:
>   • Irrigating edible gardens and container plants.
>   • Wetting compost piles.

Credit: GStibolt

The idea of water as an ornamental garden feature developed with its employment for agricultural purposes. Gardens and agriculture shared a parallel history. The terms were often interchangeable in recognition of the life-sustaining gifts provided by both landscape forms. Water quenches the thirst of both the parched soul and the dry plant.

—Chip Sullivan,
Garden and Climate, 2002

- Pre-rinsing veggies.
- Rinsing hands, feet, gloves, socks, and tools.
- Cleaning out planting containers.
- Filling birdbaths or other small water features.
- Caching enough water in a cistern or lined pond to be used for fire fighting. (See more about fire-wise landscapes in VI-7 on page 180.)

Last Thoughts

Collected stormwater can be used in many ways, all of which reduce the use of our potable water supplies. In addition, collecting and holding water on site keeps it from carrying soil, nutrients, and pollutants into storm drains, which could damage local waterways.

ACTION TOPIC III-4 — Add Water-Absorbing Features— Rain Gardens, Bioswales, and Drywells

Why This Matters

Making the whole landscape act like a sponge (called *rainscaping*) will greatly reduce the flow of runoff into storm drains, especially the most damaging first rush of storm water. This can be accomplished by adding one or more features designed to capture and/or slow down the flow rate from your property. Water-holding features like those described in the preceding pages retain water for our later use. In contrast, water-*absorbing* features help water soak *into* the soil. Both types of landscape components increase the amount of water that stays in a landscape and correspondingly reduce the amount of water that reaches storm drains and waterways.

Some states, particularly in arid areas, have water rights laws that determine whether and how much stormwater can legally be captured or harvested by an individual property owner. Check local laws to find out what is allowed in your area. In addition, many municipalities across North America have set up incentive programs to help residents install features that contribute to stormwater management. These landscape features may be increasingly important in the coming decades.

Actions

⊗ **Assess the flow of stormwater in the landscape.** To assess how your landscape handles rainwater, walk the whole property during or just after a heavy rainstorm. Notice what happens to the rain when it comes off the roof; locate where water collects and where it runs off; and determine if there is ongoing erosion. Much of this you'll already know to some degree because of the lingering results, but there are additional observations to be made during a rainstorm. After this assessment, you'll have a better idea where to add water-retention features.

You can manage the water flow by installing a combination of water-retention systems and water-absorbing features that are strung together to handle larger volumes of water. For example, you might have two connected rain barrels collecting water from a guttered roof. The overflow from the second barrel could be directed through a blind drain (see below) into a small, lined pond. The pond could overflow into a rain garden, and then the rain garden overflow could be directed to a grove of trees.

⊗ **Use blind drains and drywells to absorb rainwater.** A *blind drain* (also called a *French drain*) collects water from a downspout, rain barrel, or along a paved area and delivers it to a water-retention site. Most blind drains consist of permeable, cloth-covered pipes laid in a ditch or trench full of rocks; the pipe ends are covered with screening to keep out animals and soil. In some situations, such as a downspout or outflow from a rain barrel, it's a good idea to create a drywell to collect the water and then abut the beginning of the drain pipe near the top of the drywell (see below) so that it gathers the overflow. Be sure to build in a slight downward angle toward the outflow area.

A drywell is a hole dug into the ground that's filled with porous matter such as coarse gravel. Water occupies the gaps between rocks; a coarser fill and other porous containers, such as circles of perforated pipes or covered buckets with holes drilled into them, will create more capacity for absorbing water. Also, it can be lined with geo-textile to keep soil from washing into the space. The size of the drywell should correspond with the soil's lack of porosity—a site with poor drainage will require a larger drywell. Add soil or mulch on top of a drywell to bring it up to soil level, and then you have the option to add herbaceous plants on top to suit the site. A drywell can be added as part of a

Credit: SReed

Figure III-9: This constructed rain garden in a low spot between buildings on the University of Massachusetts campus collects water that nourishes the wetland plants. It also provides an appealing gathering spot for students and faculty.

rain garden to add extra capacity for an area with limited size and/or clay soil. A blind drain could be added near the top of the drywell to carry the overflow to the next water-retention feature.

⊙ **Add rain gardens and bioswales.** The terms rain garden, bioswale, and bio-retention swale are often used interchangeably, but some people separate them according to usage. A *bioswale* can be designed as a non-piped conveyer of stormwater from one place to another, but along the way water flow is slowed down with vegetation, curves, stones, bio-logs, or other obstacles. A *rain garden* is planted with plants which will soak up water. Good rain garden plants can withstand periods of flooding as well as drought. A rain garden can be designed as an end point for stormwater runoff, for example, from a parking area. The capacity of the rain garden should be large enough to manage water from most storms, and contain a designed overflow to direct excess water away from buildings or other infrastructure—ideally, to other gardens. For water-absorbing features, follow these general guidelines:

• Build rain gardens in natural or constructed depressions or swales. They should be designed with enough capacity to hold or absorb at least the first inch of rainfall, because the first flush of water can be the most damaging. In regions with frequent storms and high rainfall amounts, the capacity

Credit: SReed

Figure III-10: In one portion of this extended bioswale system, an elevated drainage grate provides an outlet for excess stormwater.

should be as large as possible. Capacity of the swale can be augmented by drywells built within the rain garden, or by directing rain garden overflows to additional water-retention features such as ponds, damp meadows, or forested areas.

• Locate rain gardens far enough away from buildings with basements so that saturated soil does not cause leaks into the foundations. You can build stone-lined water paths or install underground drainpipes to direct water from downspouts to rain gardens so that they are a safe distance away from buildings. Also create an overflow area, a low spot in the berm that surrounds the rain garden or blind drain, which directs excess water in the rain gardens away from buildings and into other water-retention features.

• Don't build rain gardens where they might interfere with underground infrastructure such as sewer pipes, conduits, or septic tanks and their drain-fields.

• Design rain gardens so that the water doesn't frequently sit for more than three days, to keep the mosquitoes at bay. Some water will soak into the soil and some will be absorbed by plants: larger plants with abundant leaf surface areas will remove more water from a rain garden due to their greater transpiration volume. When working in a clay soil area, build larger rain gardens or add more porosity by installing a large drywell under the entire rain garden.

• In urban areas, rain gardens can be located next to roads in median strips or sidewalk "hell" strips that have been altered with curb cuts to catch the outflows from the street. Drainage ditches can be transformed into rain gardens by adding plants, dry-wells, berms, and other obstacles to slow down the movement of water.

Note: For rain gardens in public spaces, plan for ongoing maintenance: municipal workers or a group of civic volunteers may need additional training to handle the variety of plants found in a typical rain garden. What looks like a weed to most people may actually be a desirable rain garden plant.

Credit: GStibolt

Rain garden plants can be trees, shrubs, or herbaceous plants as long as they can tolerate both flooding and drought. In areas where visibility is important, use low shrubs and herbaceous plants. When there is enough room, use trees and larger shrubs because they will soak up more water. Use well-established plants; if you purchase seeds or seedlings, plant them in containers where you can care for them until they are mature enough to handle the flooding and drought and then install them in the rain garden.

Use plants native to your region for the best success. For roadside rain gardens in areas where salt is used to de-ice in the winters, choose salt-tolerant plants. For guidance on which plants to use, survey some untended roadside ditches in your region to see which native plants are growing there. Because more extended droughts are probable in the future, it's a good idea to avoid plants that require *constant* moisture.

Figure III-11: An urban rain garden in Jacksonville, Florida, has been a team effort including the city, the St. Johns Riverkeeper, and other commercial and environmental organizations. Here, an educational sign provides information for visitors.

Last Thoughts

In an era of climate uncertainty, as both drought and intense storms become more frequent in some regions, designing all landscapes to absorb as much water as possible will help protect our plants and soil ecosystems, our urban infrastructure, and our waterways.

Why This Matters

Water features can range from small water gardens and birdbaths to ponds of any size. Larger or unlined ponds can help retain more stormwater, but any water feature will attract birds, butterflies, and other wildlife, provide important habitat, and potentially help wildlife deal with the stresses of climate change. They can be a source of water for irrigation (see more in III-2 on page 80). In addition, if a water feature is designed as part of the stormwater flow chain in a landscape, it could help to protect water quality in nearby waterways.

Actions

⟩ **Install birdbaths and trickle fountains.** Small water features can play an important part in creating habitat in small landscapes—even in balcony gardens. A simple birdbath will attract more birds if installed near a thicket of tall shrubs or a grove of trees and shrubs, where the birds have easy access to cover. Many people find that several birdbaths at different heights attract a wider range of birds and other wildlife.

To improve the habitat values of a birdbath, add a small solar-powered fountain that floats on the surface of the water. A more complex system that cycles water from a hidden reservoir through a filter and up to the birdbath will also provide a constant trickle into the water (this method avoids the need to empty the birdbath). There are attractive trickle fountains that work without the birdbath, and they, too, would add some habitat.

A somewhat more complex and expensive water feature can start with a trickle fountain or small shallow pond at ground level, which overflows into a stone-lined stream bed that flows downhill to a lined pond, where a submerged pump then cycles the water back to the top again. This type of water feature provides a good focal point in the landscape and excellent habitat for a wide variety of birds, frogs, butterflies, and other wildlife.

⟩ **Create water gardens.** You can build a stand-alone water garden in a barrel or in a tub sunk into the ground where you can grow emergent and/or floating plants. It's best to add a few small fish to prey on mosquito larvae. Depending on the climate, frogs and other animals may take up residence, as well. As with birdbaths, a trickle fountain will keep the water moving.

● **Build or install a pond.** A pond and its moisture-loving plants provide excellent habitat that will attract birds, dragonflies, turtles, frogs, and more. Mosquitoes tend not to be a problem around a pond because of all those insect predators. Ponds can range in size from small preformed tubs set in the ground to large groundwater or spring-fed bodies of water.

If you are not blessed with a natural pond, it might be worthwhile to add one for the amazingly rich habitat it creates. Use moisture-loving native plants around the edges and emergent plants in the water. (Emergent plants are those that grow in shallow water with their tops growing above the water level.) If possible, create a damp shoreline where butterflies can mud puddle to consume water and minerals.

Figure III-12: You can manage the flow of stormwater from collection devices—such as rain barrels and lined ponds—to absorption features like rain gardens and planned-for wet places.

Note: It's not a good idea to stock any outdoor pond with koi (carp) because they are known to be invasive and harmful to natural aquatic ecosystems if they escape. Always try to use fish native to your own region.

In arid climates, it may be difficult to sustainably manage a pond in the landscape; the usual members of a pond ecosystem may not find your pond. However, if there is a natural oasis in your region, use it as an example of how to set up your own private oasis. As discussed in the Overview to this Section, one or more ponds can be part of a rainwater flow plan, with larger ponds serving as endpoints for stormwater flows through the landscape. As with rain gardens, locate ponds far enough away from buildings with basements so that saturated soils do not cause leaks.

Figure III-13: Garden in the Woods, a 45-acre preserve in Framingham, MA, includes a natural pond with artificial floating islands that provide habitat for aquatic wildlife, including the one basking turtle shown here.

Credit: S Reed

⊙ **General guidelines for lined or artificial ponds:** If you wish to create a new pond in a sandy substrate with no immediate groundwater to support the water level, then you'll need to install a liner of some sort. These general guidelines may be helpful.

- Unless using a preformed tub as a liner, create as big a pond as space and budget allow. It's much easier to dig the hole and lay the liner all at one time. Be sure to plan the overflows so they direct water away from buildings and into an area that can absorb any excess.

- If at all possible, use rain barrel water or diverted roof runoff to fill the pond instead of potable tapwater. Also, if possible, inoculate the water with a bucket of water from a natural pond or lake to introduce beneficial microbes. After it's filled, start planting with your potted plants of rushes, irises, sedges, and other regional emergent plants.

- For potted plants submerged below water, cover the top of the soil in each pot with weed barrier cloth held in place by a layer of sand or gravel.

⊙ **Create natural swimming ponds.** A natural swimming pond can be an appealing alternative to a traditional chlorine pool. It consists of a swimmable body of water connected, either directly on the surface or via underground piping, to an area of shallow water where wetland plants are growing. These plants, specially selected for this purpose, take up nutrients from the water, thereby depriving algae of the food they need to survive. The result, when all goes according to plan, is clean, clear water. In many such ponds, an extra biological filtration system, customized for the size and location of the project, further helps keep the water clear.

Natural pools are climate-wise. From dragonflies that gobble up mosquitoes, to frogs and toads plopping among the lily pads, to wild herons and kingfishers that dine in this miniature ecosystem, the variety of species potentially supported by a natural swimming pond provides tremendous value, both for our own enjoyment and for the health of the environment. In addition, such ponds (like any body of water) can help cool the air and increase our outdoor comfort during hot summers.

Credit: SReed; Designer: Waterhouse Pools

Figure III-14: This natural swimming pond was constructed with an attached shallow-water vegetation area, where the plants act as a filtration system for the swimming water.

⟩ **Adopt a water-retention pond.** In some wet regions, retention ponds are mandated to collect stormwater runoff from roads or parking lots. Unfortunately, ongoing care for these water features may be a low priority for municipalities, but retention ponds can be transformed from an eyesore into an attractive neighborhood asset.

Before taking any action, check with local government about any regulations that apply. There are many ways to improve the health, appearance, and habitat values of a neighborhood retention pond. Improvement projects could include repairing eroding banks, reduction of fertilizer runoff into the pond, trash removal, invasive plant removal, and native wetland plant installation.

It will probably take several years to restore a neglected retention pond, but as a water feature in the community, it does offer a great wildlife habitat and an important cooling feature in a residential neighborhood or a commercial park. In any case, community-wide effort and increased awareness of the pond

Figure III-15:
Swamp mallows and
horsetails populate
a wet section of the
High Line Park in
New York City.

Credit: GStibolt

may motivate similar efforts beyond this retention pond's watershed. *Note*: Be aware that many towns and cities limit access to all open water bodies, to prevent accidents.

Last Thoughts

Water features, even small ones, can serve as important focal points in the landscape for both humans and wildlife. They also allow for somewhat more diversity by supporting plants that require moisture, which otherwise might not occur in a conventional or drought-tolerant landscape.

Additional Resources

- epa.gov/watersense
- water.usgs.gov/education.html
- *Water-Saving Garden: How to Grow a Gorgeous Garden with a Lot Less Water*. Pam Penick. Ten Speed Press, 2016.
- Ginny's *Green Gardening Matters* blog includes links to articles on rain gardens, rain barrels, and drywells on its References page. GreenGardeningMatters.com

may motivate similar efforts beyond this retention pond's watershed. *Note*: Be aware that many towns and cities limit access to all open water bodies, to prevent accidents.

Last Thoughts

Water features, even small ones, can serve as important focal points in the landscape for both humans and wildlife. They also allow for somewhat more diversity by supporting plants that require moisture, which otherwise might not occur in a conventional or drought-tolerant landscape.

Additional Resources

- epa.gov/watersense
- water.usgs.gov/education.html
- *Water-Saving Garden: How to Grow a Gorgeous Garden with a Lot Less Water*. Pam Penick. Ten Speed Press, 2016.
- Ginny's *Green Gardening Matters* blog includes links to articles on rain gardens, rain barrels, and drywells on its References page. GreenGardeningMatters.com

ECOSYSTEMS

Overview

Everything alive is performing some kind of role in nature. From the topmost leaves of a maple tree to the sap that rises from its roots…from the red-tail hawk soaring far above a meadow to the grass-spider seeking prey amidst the stems…from pine pollen dusting a breeze to peony pollen on the knees of a bee…every plant and animal is doing *something* that affects other plants and animals. And everything they are doing, whatever it might be, is integral to the working of the natural world.

We humans might notice some of what's going on, especially when the sight is pleasing in some way, or dramatically askew. But quite often, especially when things are tiny or hidden, or just proceeding in the normal course of natural events, we don't see them at all. And most of the time, most of us don't understand or truly appreciate the significance of nature's countless activities. Yet the elegant interplay of species is what makes life possible.

Ecosystems Services

Collectively termed "ecosystem services," the value to society provided by healthy, functioning ecosystems has become increasingly recognized since the 2005 publication of the United Nations' "Millennium Ecosystem Assessment," which articulated four categories of such services:

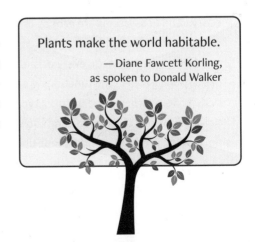

Plants make the world habitable.

— Diane Fawcett Korling,
as spoken to Donald Walker

Figure IV-1: Native to eastern North America, blue flag iris thrives in wet meadows, marshes, and along streambanks. Its beauty easily catches our eye. But do you see the little bug on its petal? A diversity of insects is vital to the working of nature.

Credit: SReed

- Provisioning: providing food, medicine, water, and habitat.
- Regulating: controlling climate, erosion, flooding, disease, etc.
- Supporting: enabling nature to work via nutrient cycles, pollination, etc.
- Cultural: providing spiritual and recreational benefits.

These services are often viewed in terms of how ecosystems help people. And they do. But it's important to understand that we human beings are the incidental beneficiaries of the complex working of ecosystems. Our lives depend on them, but we should not make the mistake of thinking they are there *for* us. Ecosystems are both the cause and the result of natural functioning. Ideally, we should play a supporting role. Beyond enjoying all the

What Are Ecosystems?

Habitat is the physical place where an organism lives. A group of organisms of the same species, living in the same habitat at the same time, is called a *population*. Populations of different species living together make up a *community*.

With the addition of the physical, non-living elements, a community becomes an *ecosystem*. Ecosystems are groups of organisms plus their habitat.

Within an ecosystem, every organism has a niche: a role in the functioning of that system. Sometimes niche means a type of service provided by an organism, such as pollination, predation, or decomposition.

Sometimes niche refers to a physical location in the habitat, such as in the tree canopy, or at the edge between two types of habitat, or in the leaf litter.

Biodiversity happens when a region contains many habitats and ecosystems, and/or when an ecosystem contains many niches occupied by many different species doing many different jobs.

Note: Ecosystems are ecological phenomena. They differ from plant hardiness zones, which are a gardening/horticulture concept. (Read more about hardiness zones in II-2 on page 58.)

fruits of nature's labors, beyond taking what we want, we also ought to give back to nature in some way. And although ecosystems don't depend on us to exist, they do need us not to ultimately destroy them.

We human beings have been altering our surroundings since the dawn of time. Our presence has sometimes been benign, but often we have diminished the wellbeing and survival of other creatures. Granted, in the past century or so we have protected large swaths of land: about 14% of the US and 11% of Canada are preserved under federal, state, tribal, provincial, or local authority.[1] But we have also converted hundreds of millions of acres of ecologically functional land into mostly lifeless subdivisions, corporate campuses, and urban sprawl. Even now, barren lawn carpets more than 40 million acres of the continental US. Natural areas throughout North America are shrinking and getting fractured into smaller and smaller patches that are often separated by miles of unlivable, or uncross-able, human-dominated landscape.

And, now, with human assistance, climate change is making matters worse—with a speed that can outpace a species' ability to adjust, migrate, or evolve. As a result, many ecosystems are experiencing stresses that make them particularly vulnerable to decline or collapse. Climate change raises the stakes and increases the need for us to take care of the environment even more than before.

In This Section

Fortunately, we can make things better. Every yard is part of the larger fabric of the region in which it exists. Every landscape is an integral swatch of the whole beautiful tapestry of nature. And almost everything we do on our own property has the potential to help the larger environment.

Credit: SReed

Figure IV-2: Pokeweed grows in clearings and woodland edges across the eastern US and Canada. The plant can be medicinal or toxic for humans, but its midsummer flowers and late summer berries provide important food for pollinators and songbirds.

> Humans cannot live as the only species on this planet, because it is other species that create the ecosystem services essential to us. Despite the disdain with which we have treated it in the past, biodiversity is not optional.
>
> —Doug Tallamy, *Bringing Nature Home*, 2007

Figure IV-3: Wild turkeys prefer open, mature forest as habitat, but they seem able to thrive in almost any native plant community that includes cover and clearings, such as this patch of lawn surrounded by woods.

Credit: SReed

The following *Action Topics* show how we can contribute to the solution:
- IV-1: Prioritize for Native Plants
- IV-2: Plant in Groups and Communities
- IV-3: Create Habitat-Rich Layers and Edges
- IV-4: Enhance Biodiversity
- IV-5: Create Semi-Wild Patches

Many outstanding native plants have been in use now for a long time. And demand for them is increasing. They fit into the informal, intimate, seemingly unstudied effects that are sought for in many grounds and gardens today, where flowers are luxuriantly intermingled, boundaries are freely planted, trees are irregularly grouped and lawns are sometimes left unclipped. More than this, the very landscape of which they are so vital a part is beginning to be retained and recreated as a setting for the house.

—Edith Roberts and Elsa Rehmann, *American Plants for American Gardens*, 1929

A PRIMER ON...

Native Plants

What Is a Native Plant?

When the ecological landscaping movement began in the 1920s, "native" simply meant a plant that grew naturally in a particular location in North America before European colonizers introduced plants from elsewhere. This general definition remained unchallenged (and largely unnoticed) for decades, as landscaping fell from importance during the Depression and Second World War. The idea faded even further with the post-war rise of the lawn and a modern landscaping style that treated plants mainly as spatial and sculptural elements.

However, after the 1962 publication of Rachel Carson's *Silent Spring* and the first Earth Day in April 1970, public interest in the environment reawakened. And with that interest came probing questions. How can we exactly identify which plants are native and which are not? Are past botanic records actually complete and reliable? How do we set a colonization date for every region on the continent (and what about the rest of the world)? And isn't the date of European settlement itself an arbitrary cutoff point?

Questions like these have gradually inspired a re-thinking of the concept of "native." Geographic and chronologic delineations no longer quite capture the true meaning of a term that was originally understood to include ecological realities and processes. Now, nearly a century after the term "native plant" came into use, here we propose a more mature and up-to-date definition:

Native plants are species that have no known history of importation and have lived in a place long enough to develop the particular relationships with other species that constitute the working of nature.

Credit: SReed

Figure IV-4: Like most butterflies, this adult black swallowtail feeds on a variety of plants, including this mountain laurel, but its larvae will eat only certain plants in the carrot family, including dill, parsley, and Queen Anne's lace.

What's Good about Native Plants?

If properly chosen, and once established, native plants are generally able to exist without much input of resources or outside energy. Their soil acidity requirement is often—though not always—easily accommodated in soil that exists on site. (But be aware: this will need to be verified in each case, because often the soil in our yards has been dramatically altered.) And because native species are adapted to local conditions, they usually need less water during dry times, which preserves that precious resource and reduces pressure

Credit: G. Stibolt

Figure IV-5: The Seminole pumpkin (*Cucurbita moschata*) originated in South America and was traded among the indigenous peoples northward into Florida. However, although it predated the arrival of the Europeans, it grew only under cultivation, not in the wild, and so is not considered a Florida native plant.

on water infrastructure such as pipes and sewers. *Important caveat*: Native plants are still plants; when used in our domesticated landscapes, they will need care and usually extra watering to become established, and even once established they may still require some attention.

Beyond these well-known attributes, though, is native plants' role in supporting ecosystem health. We might even suggest that this is their most valuable quality. Why? Because native plants provide food that is edible and useful to native insects, and insects constitute the protein that many birds and mammals need to feed their young. Many insects are "specialists" who can get nutrition from only one or a few species of plants. As a result, they depend entirely on the plants with which they have co-evolved. And these are usually native species.

Furthermore, many native plants provide food and habitat for an immense variety of species. For example: oak trees are nibbled on by the caterpillars (larvae) of more than 500 species of moths and butterflies, plus they supply food, cover, and nesting sites for more than 250 species of birds and mammals. Non-native plants may also provide edible fruit and pollen, but in general they support far fewer species than do native plants, and they also have the potential to become invasive.

ACTION TOPIC IV-1 Prioritize for Native Plants

Why This Matters

In recent decades, human-dominated landscapes have taken over more and more natural habitat. Parks and conservation lands are no longer sufficient to sustain displaced wildlife and stressed species. So the idea of people supporting nature and ecosystems has become increasingly important. A good way to help is by planting as many native plants as possible in our domestic and managed landscapes. ("A Primer on Native Plants" explains why.)

Actions

▶ **Identify plants that are native to your region.** The USDA provides a comprehensive nationwide list at www.plants.usda.gov. You could even go further and check with local nature centers, university extension services, local herbaria, or native plant societies (www.plantsocieties.org). Online lists can be a good source, as they tend to be more up-to-date than published lists, especially with regard to native ranges. But it's always a good idea to check two or more sources, just to be sure.

Credit: SReed

Figure IV-6: Just a few species from an eastern US native plant community—including witch hazel, mountain laurel, mountain andromeda, paxistima, and Christmas fern—provide a rich adornment to this lawn-free, woodland landscape.

Note about species ranges: Northward shifts in range happen very slowly, especially so with plants, and they are very difficult to predict. (To learn more about *assisted migration* see the sidebar in VIII-6 on page 242.)

⊙ **Buy native plants,** starting at local nurseries and garden centers. It's tempting to buy plants marked "native" at big box stores, but these labels aren't always accurate and sometimes represent more of a marketing strategy than botanical reality. So, if possible, no matter where you buy, take a moment to verify.

⊙ **Request native plants anywhere plants are sold.** This will help signal to retailers that there's a viable market for this type of plant.

⊙ **Visit and study natural areas.** If possible, observe which plants grow where, how they look and how they arrange themselves. A good plant identification book will help, but careful observation may give you enough information and inspiration to then find some plants at a nursery. Master Gardeners at your local agricultural extension offices can help with identification.

Figure IV-7: Grow Native Massachusetts is one of the growing number of organizations dedicated to promoting the beauty and value of ecological gardening, using regionally indigenous species.

Credit: SReed

Credit: SReed

Decide how to use the native plants. Once you know which plants you want, think about how you will incorporate them into your landscape. Consider these options:

- *Substitute natives for conventional landscape plants.* A highbush blueberry, with its edible fruit and brilliant red fall foliage, is a great replacement for burning bush. The latter does provide food for some birds and butterfly larvae, although probably less than the native blueberry, but it's invasive in many areas; it tends to be aggressive and destructive of native ecosystems.

- *Create groups and assemblages of plants that are inspired by naturally evolved associations.* (IV-2 covers this subject in more detail.) Instead of planting a single maple tree in the lawn, think about creating a small grove of trees, all growing as they might in the woods, with their roots and branches intertwining and the large swath of ground below the trees mulched with natural leaf fall.

- A good way to learn which plants naturally congregate is *to attend native plant conferences or workshops.* Or check books or online resources appropriate for your area. See E.G. Gary Hightshoe's *Native Trees, Shrubs and Vines for Urban and Rural America* for it's amazingly thorough information about native plant preferences and associates.

Figure IV-8: Native highbush blueberry (*left*) can provide similar red fall foliage as the more familiar but invasive burning bush. And the blueberry supports a wider variety of pollinators and wildlife without endangering natural plant communities... plus, we can enjoy the fruit!

- *Create approximate versions of native ecosystems* that you observe in your region. In general, this process begins with some study to learn which plants want to grow there, their natural patterns of distribution, and the conditions they need to get started. Sometimes you can make this an extension of a wood's edge, or simply add to an existing small tree or hedge. This is one of the times when getting the advice of an experienced gardener or landscape designer trained in ecological design can be very helpful.
- *Cultivate a new way of thinking.* Instead of saying, "I'll try to use some natives, if possible," how about approaching every garden decision with the question, "Is there some very compelling reason not to use a native plant here?" In every part of every garden and landscape, we could seek to plant a native first, and choose a non-native only when there's no native that could possibly work in its place. Or just change the design of the landscape to not need that non-native plant at all.

Working with native plants is such a joy. The more you use them in the landscape, the more alive it becomes—with birds, butterflies, bees and pollinators—and even amphibians and reptiles if we are lucky, and if we do enough to create healthy plant communities and ecosystems.

I really want the American toad to find a home in my garden.

— Claudia Thompson, Founder, Grow Native Massachusetts

Last Thoughts

In nature, when species live together for a long time, they co-evolve. Whether fern, fungus, sparrow, or newt, plants and animals in a shared ecosystem continually influence each other in ways that might be perfectly obvious, beautifully subtle, or completely unseen and unknowable. When one sort of plant suddenly appears in the midst of other creatures it has never encountered before, and who have never encountered it, all bets are off. The services it used to provide may or may not be useful in its new home. The bugs and caterpillars that used to dine on its leaves might or might not recognize it as food. Birds may or may not appreciate the seeds or fruit it produces at a strange time of year. This new plant might be of no benefit at all to its neighbors, or it might actually be a detriment, potentially overwhelming the whole neighborhood, taking up more and more space in the absence of the predators, weather extremes and diseases that originally kept it in check. [Excerpt from *Energy-Wise Landscape Design*, by Sue Reed]

ACTION TOPIC (IV-2) Plant in Groups and Communities »

Credit: SReed

Figure IV-9: A flowering dogwood blooms at the edge of this front yard grove of trees, just as it would in nature as a woodland understory tree.

Why This Matters

Planting in groups, groves, clusters, and assemblages will benefit the environment—and the gardener—in several ways. First, this type of little ecosystem will provide a variety of niches where many species can do their work and thrive. The creatures that visit one plant can interact with other plants too, and with other creatures. And ideally this will create a diverse and complex web of life, which will enhance the health, vitality, and resilience of the whole planting.

Second, the soil in which all the plants are growing can be treated like one entire ecosystem. The natural fallen leaves can

Moving from the idea of traditional plantings to a designed plant community starts with letting go of the idea of plants as objects to be placed, like pieces of furniture. Instead, think about plants as groups of compatible species that interact with each other and with the site. When plants are paired with compatible species, the aesthetic and functional benefits are multiplied, and the plants are overall healthier.

— Thomas Rainer and Claudia West, *Planting in a Post-Wild World*, 2015

stay in place and return nutrients to the soil as they break down. The mulched surface will help moisture seep in and be shared between many plants. And the roots of all the plants can interact in their invisible chemical relationships (some in cooperation, some in competition) that make up the dynamic work of nature. Further, plants growing in groups tend to be more drought-tolerant and wind-resistant than single individuals.

Finally, planting in groups and groves will help with storing carbon. By creating beds where a large swath of soil remains undisturbed and gets gradually richer and richer, more carbon gets stored in the soil (see II-3 on page 63 for an explanation of how this works, via photosynthesis). And instead of planting an isolated tree or shrub that might be easy to abandon later if we have a change of heart, planting a whole grove or island or peninsula makes a commitment to all the plants. And this increases the likelihood that their stored carbon will remain safely out of the atmosphere.

Figure IV-10: Steeplebush and purple coneflower make as congenial a group here, in a New England residential landscape, as they do in wild meadows.

Actions

⊙ **Choose an area that seems likely to remain undisturbed** for several decades. Plan for the maximum width of branching, and avoid planting under/near power lines, which might require later pruning or removal of the trees. Also avoid underground infrastructure such as gas, water, or sewer pipes, septic systems, irrigation system pipes, or electric and cable lines. If you don't know what's underground, you can call 811, a free service across the US that puts you in contact with the local utility companies that will send out representatives to mark their lines.

If established woods or other ecosystem already exists on the property, consider creating either an extension or a peninsula to enlarge the overall size of woodland habitat. Remember, though, that our human developments have already created plenty of edge

Credit: SReed

habitat, so it's preferable, if possible, to simply enlarge the outline of a woods, thereby increasing the amount of interior habitat, which many species need.

❯ **Identify the tree species that already exist** either within the property, or in nearby/local natural areas. These are the plants most likely to tolerate and thrive in the conditions of your property. Look for the less common trees and shrubs in these woodlands and plant some of those in addition to the dominant or most-common species. This will help increase the local diversity. Exceptions to this advice include:
  • If your own property has distinctly different soil or bedrock conditions than are typical in your region, this will necessitate further research to match species (consult with nature centers and botanic gardens for advice).
  • If the location of your intended grove is in a distinctly different microclimate than other nearby woods, i.e., a steep shady ravine, an exposed west-facing slope, etc., these conditions will also necessitate added research into which species will be appropriate.

Include a mix of species, including tall trees, understory trees/shrubs, and ground-layer vegetation.

❯ **Repair scarred edges.** If a woodland edge has been chopped away, perhaps to create an opening for construction or just to expand a yard area, seek out and plant large and small (understory) trees and shrubs. Leave space for (and adjust expectations to tolerate) some scruffy growth, as young trees mature. Encourage ferns, grasses, and shade-tolerant wildflowers to colonize the ground at the edge.

❯ **Prepare for future tree planting.** If the soil is particularly barren, dry, or sandy, spread leaves/compost/mulch over the ground and leave this layer to decompose for a year or so to encourage the start of microorganisms and fungi that will help tree roots grow. (Read more about mulch in V-5.) If you just wait a while, some volunteers might show up that you decide you'd like to keep; plants that "plant themselves" are likely to be well adapted to their conditions. You can edit out the ones you decide you don't want.

❯ **Plant your selected species in distributions and patterns** that resemble the way they arrange themselves in nature. For trees, include younger and older individuals. Plant some individuals closer together than you would if they were to be fully rounded specimen trees. Let their branches intermingle, as they do in the woods; the plants themselves will work out the space they need and the space they can't share.

❯ If the site is very sunny, **consider planting some fast-growing, shorter-lived species** to provide quick shade for slower, longer-lived trees that could eventually dominate. This may mean removing those "nurse" plants at some point if they begin to interfere with the growth of preferred species. Or it could mean just letting nature work it out.

Last Thoughts

One way for landscapes to be healthy and robust is for us to intentionally arrange plants in groups, so that all the life that's happening belowground (roots, microorganisms, shared chemistries, etc.) and all the life that's happening aboveground (pollinators, insects, birds, chipmunks, etc.) can freely interact, compete, cooperate, grow, share, eat, die, and decompose in whatever ways nature figures out will work the best. Planting things in groups and groves helps make this happen.

> When the area including your garden site becomes a habitat or even a series of related habitats, that means you have begun to understand that the conditions found there dictate what plants and animals will thrive in your garden. And when, instead of a collection, the plant inhabitants of your garden become a community, you are beginning to understand that the species that live in a habitat are not just accidental neighbors but that they by definition tolerate the same conditions, and in some cases they may interact in a positive way.
>
> —Larry Weaner and Thomas Christopher, *Garden Revolution: How Our Landscapes Can Be a Source of Environmental Change*, 2016

ACTION TOPIC (IV-3) Create Habitat-Rich Layers and Edges

Why This Matters

Layering plants is not a new idea in garden design. Here, however, we take the idea beyond just how the garden looks to how vertical layers also help support wildlife and diversity, which will help our landscapes better adapt to stresses and disruption.

All natural ecosystems consist of structural layers, or what some call the *architecture* of nature. For example, in the eastern US and other temperate regions, a typical forest community includes (in addition to species living in the soil) plants near the ground surface, shrubs and saplings, and canopy trees that might tower 150 feet above the floor. Prairie, grassland, and desert communities also contain layers, but the layers may come into being over time, as a growing season progresses. Even a mountain bald can support multiple layers of plants, from lichens skinning the rock's surface to gradually taller and taller (but still relatively tiny) alpine vegetation.

What's important here is that all of these layers provide habitat and specialized niches for a wide variety of animals to live, hunt, forage, and (in some cases) reproduce. Some animals might live in very thin slices of nature, while other species might range across many levels. Some migrating species need to

Credit: SReed

Figure IV-11: From the Jack-in-the-pulpit and ferns near the ground, to the hazelnut shrub just above, to the maple canopy over all, this little community offers several layers of habitat.

occupy a certain layer of habitat only temporarily as they pass through from south to north and back again. Overall, it is this network of countless individual niches that makes the whole system work. So the more levels we provide, the more species of wildlife we support.

Similarly, ecosystems usually contain a mosaic of several communities, interweaving and bumping against each other based on plants' abilities to thrive in various soil, moisture, and light conditions. The intersection where one community changes into another is a transition zone called an *ecotone*. Whether narrow or wide, gradual or abrupt, local or regional, this dynamic edge can be loaded with opportunity for many species to thrive because the zone where communities overlap contains more niches than the individual communities on either side, and therefore can support species common to both. The resulting increase in diversity is called the *edge effect*. We can create these rich edges in our yards and gardens, and the resulting landscapes will be more resilient than they would be without them.

Note: Keep in mind that the natural world needs more deep interiors than it needs edges and fractured or diminished ecosystems. However, if we are going to create edges, we should make them dynamic edges full of layers, complexity, and diversity.

Figure IV-12: Between wildflower meadow and maple woodland, a transition zone (ecotone) of saplings and shrubs offers cover and habitat for species passing back and forth.

Credit: SReed

Actions

If your landscape is older and established, take a close look at the plants that are already growing there. Notice which ones are doing an important job such as shading the house, driveway, or patio, screening or framing a view, providing rich habitat for wildlife, or serving as the start of a possible new community. Preserve those. Notice which are thriving and which are struggling.

If dealing with new construction and a bare yard, start out by defining the location and shape of desired outdoor gathering spaces, destinations, and the

pathways that will connect them, and decide on the places for ornamental/food gardens. If you plan to include rain gardens or bioswales to manage stormwater runoff, note their probable locations.

To Create More Layers

- ❯ **Develop a list** of which and how many plants to add by consulting resources that list plants according to their natural communities and associations. Resources that list plants simply in terms of how they look together or in terms of which growing conditions they require miss the point: what matters is how *plants arrange themselves* in communities of their own choosing. We might not understand the reasons, but they work for the plants. If the naturally evolved communities contain a particular assemblage of plants, it's reasonable to assume that those plants will also thrive together in similar conditions in our yards and landscapes. See "A Primer on Native Plants" on page 103 for more details and resources.

- ❯ **Remember that layering takes time,** so if possible start by planting the species that will grow the largest, to help the new ecosystem get started as quickly as possible. When planting trees, choose smaller, younger plants grown in containers over larger plants dug from the ground and "balled and burlapped." Older trees take much longer to adjust to their surroundings than younger ones, and this can delay their growth to a point where the younger ones actually sometimes catch up in size. Plus the mortality rate after planting for larger trees is higher, so whenever possible it's better to invest in younger trees.

Figure IV-13: F formal or mar assemblage c contain layers found in natu such as this g mountain lau sheep laurel.

Credit: SReed

To Add More Complex Edge Habitat

>> **Add habitat at the boundary** where lawn meets an established grove or woodland. Either dig a hole and plant through the turf, or mulch the lawn grass in the targeted area. Either way, use mulch around, but not touching the new plants. This is least energy-consuming, and it avoids disrupting the soil, which preserves soil structure and carbon and eliminates the possibility of erosion. Or if you prefer to remove existing lawn/turf in the area, see Section I for information about removing lawn.

>> **Treat the soil as a whole system.** We know that planting trees in big holes filled with amended soil doesn't produce good results. It's a bit like putting a tree in a huge pot: sometimes the roots just circle round and round in the hole.

Figure IV-14: In June, the three visible layers of this little grove include striped maples up high (with their dramatic dangling samara seeds), azaleas, and bracken ferns in the middle, and wild sarsaparilla down low. Not easily seen are the sarsaparilla flowers and seed-heads (shown in the close-up) tucked under the umbrella of foliage, along with a carpet of native pachysandra (Allegheny spurge), whose flowers bloomed and faded long before, in the light of early spring.

A better solution is to create the optimal soil before any planting: spread a thick layer of leaf litter—ideally collected from within the property—and let it decay for about a year. This gives the beneficial soil organisms time to mix the nutrients downward.

After a tree is planted, add a top-dressing of locally made compost outside the planting hole to entice the tree roots to grow outward. You can do this at the beginning of each growth season; each time, place the compost farther away from the tree. Wide-spreading roots make the tree more wind- and drought-tolerant. (See V-4 on page 144 for more about compost.)

> What is good may not look good, and what looks good may not be good. The difference between the scientific concept of ecology and the cultural concept of nature, the difference between function and appearance, demonstrates that applied landscape ecology is essentially a design problem. [The solution] requires placing unfamiliar forms inside familiar, attractive packages. It requires designing orderly frames for messy ecosystems.
>
> —Joan Iverson Nassauer,
> *Landscape Journal,* 1995

Last Thoughts

To some people, edges that have been "diversified" appear less tidy and tended than the more typical abrupt edge between woods and lawn. To relieve any problems caused by this new approach, create crisp and tidy boundaries between lawn (if any exists) and the new softer edge. Or, as with meadow discussed in I-6, add *cues to care,* i.e., elements that signify this new landscape is intentional, and is being actively tended.

ACTION TOPIC ((IV-4)) **Enhance Biodiversity** »

Why This Matters

As mentioned in the Overview to this Section, we human beings depend on having a functioning natural world that produces oxygen, cleans the air and water, and cycles nutrients. These ecosystem services benefit all of us directly. Less tangible effects, like providing beauty and pleasure, inspiring creativity, and soothing the soul, also make nature essential to our lives.

And now we also recognize the value in natural areas' ability to help mitigate climate change by absorbing and storing carbon, cooling the air, and pulling water from the ground to the air, thus generating rain and weather. (Also, let's not forget that other species have their own *inherent* value, apart from any "service" they may provide to us.) All of these gifts from the natural world help human beings survive and thrive. But what enables the natural world to survive and thrive?

The answer is simple: biodiversity. Biodiversity means that many different species are taking advantage of the available resources and producing useful biomass, food, and various other substances that organisms in the ecosystem can use. It means many species are competing with, preying on, feeding from, and sustaining other species. It means that many organisms are doing overlapping jobs within the system, with the result that if one group or species struggles, or if one population declines, there's likely to be other species available to pick up the slack and keep the whole system going.

This overlap of jobs tends to make ecosystems resilient in the face of dis-

Figure IV-15: Beneath this sugar maple, what used to be thin, scraggly lawn is now a little ecosystem full of ferns, small native honeysuckles, sprawling laurels and various herbaceous plants that show up one year and disappear the next. Management is minimal here, and natural processes rule. (*Note:* The "window" between canopy and ground creates a wind tunnel effect that helps cool the landscape.)

Credit: SReed

ruption and disturbance. And it means that even if the makeup of an ecosystem changes somewhat over time, its overall functioning endures. So, for example, a bio-diverse forest will continue to cool the air, produce oxygen, store carbon, absorb rainwater, transpire moisture to generate more rain, and support abundant wildlife even if one species is in peril. A diverse riverside ecosystem will continue to prevent soil erosion, filter pollutants, shade and cool the water, and support a variety of wildlife.

What Is Biodiversity?

There are three types of biodiversity that combine to form the variation and complexity of life upon which nature's functioning depends.

- Ecosystem diversity can help a region recover from disturbance. This happens because one ecosystem may hold an assortment of species that have the potential to re-colonize other ecosystems when they get stressed by human-caused disruptions or natural events such as fire, flood, hurricane, large-scale die-off, or migrations. All of these stressors are likely to become more common as a result of climate change.
- Species diversity within any individual ecosystem makes that ecosystem more productive, stable, and resilient, and therefore more able to adjust to new conditions, such as drought or changing weather patterns. This happens because with more species come a greater potential for redundancy, so that if one species struggles or fails, its function within the ecosystem—for example to provide certain food nutrients, a specific type of shelter, or an essential chemical interaction, particularly in soil ecosystems—may be performed by one or more other species.
- Genetic diversity, i.e., the variety of genetic material within a local gene pool, helps individuals produce—via sexual reproduction—new generations of individuals (seedlings) who will be variously more or less able to survive new conditions and, when things work out well, to procreate further. This is the essence of natural selection, which in plants depends on cross-pollination and the production of seeds. And, along with ecosystem and species diversity, it is an essential component of survival in this new era of climate uncertainty and change.

Actions

Environmental conditions—the availability of resources, the effects of disruptions and invasions—combine to create the character of a natural area and determine the variety of habitats and species that can exist in an area. And it is this biodiversity that determines nature's capacity to function, adapt to change, and endure. So, if we want to support the natural world—and, by extension, our own survival—we need to support biodiversity. Here are some steps we can take toward that end:

- **Plant a rich variety of species,** ideally those grown locally, using locally adapted seed or stock.

- **Use a variety of flowering and fruiting plants** to support pollinators, birds, and small mammals.

- **Plant primarily native species.** Hybrids and cultivars (and yes, even *nativars,* which are cultivars of native species) are created by endlessly replicating the same preferred attributes such as flower size, form, or color, or the size/shape of foliage, or other qualities that are valued by gardeners. But these qualities might be either useless or detrimental to other species that rely on those plants for nectar, pollen, nutrients in the foliage, and/or nesting habitat material. By using mostly native species, especially those produced from seed (genetic reproduction), we can help promote genetic diversity. This is natural selection, an essential engine that drives evolution. And a species' ability to evolve will become even more necessary as conditions change and organisms need to adapt in every way available to them.

- **Make your landscape into a mini-ecosystem.** Model some portion of your yard or gardens on one of the local plant communities or associations found either in natural areas or those on the undeveloped portions of the site.

> To achieve high levels of genetic diversity, it is important that plants be propagated by seed, rather than vegetative methods, to allow sexual out-crossing and the accompanying genetic shuffling to occur.
>
> —Travis Beck, *Principles of Ecological Landscape Design,* 2013

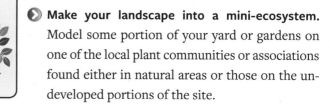

▶ **Plant for pollinators.** Get acquainted with more insects in your yard; identify and recognize them. Realize that many insects are specifically designed, as the result of co-evolution with native plant species, to require certain vegetation and sometimes no other. Appreciate their value as the base of the food chain that supports the birds we want to invite into our yards. (Read VII-2 for more about gardening for pollinators.)

▶ **Encourage diversity in soil ecosystems.** Minimize soil compaction, build soil structure with compost, minimize soil disturbance (i.e., digging and turning), use amendments that encourage soil microbes, and minimize the use of nitrogen-based fertilizers (See Section V for more about soil ecosystems.)

Credit: S Reed

Figures IV-16: A few years after the mowing stopped, this liberated lawn sprouted its own community of common milkweed, where Monarch caterpillars dine and other pollinators thrive. Phoebes, which nest in the author's nearby porch, raise their babies here year after year, by darting back and forth to this handy cafeteria.

Removing Invasive Woody Plants

This subject is extremely complex. Sometimes just cutting plants down will be enough to prevent further spread, while in other cases this might encourage roots to grow even more vigorously in countless new directions. Sometimes digging out the entire root mass can be effective, but other times not. This book is not the place for detailed instruction on this subject. For guidance about best strategies for each type of invasion, check with a local department of natural resources, regional university extension service, botanical garden, nature center, the USDA, or—ideally—state and regional native plant societies.

> You don't need to tear up your yard to have a more biodiverse landscape. You can just look around your yard and find the places where you have mulch or gravel, and start thinking about what plants you could put there with other existing plants.
>
> — Thomas Rainer, video: *Planting in a Post-Wild World*

❯ **Keep an eye out for invasive plants** and remove them, ideally as soon as you know they're there. Eliminating them when they're small and just getting started is a lot easier and more effective than waiting until they're well established and widespread. *Remember*: Gardening for biodiversity is not the same as doing nothing. (See more about invasive plants, including definitions, in VII-3 on page 210.)

Last Thoughts

Ecologists debate the countless small details about how biodiversity works to support ecosystem health. The scientific community is not yet completely sure how to quantify all the large and small effects. No study says precisely how much diversity is needed. But all agree on this: greater biodiversity tends to be the primary and most reliable indicator of ecosystem stability, resilience, and long-term vitality. As such, it is a quality that's always worth nurturing—in landscapes both wild and tamed. And this is more important now than ever before.

> Leaders in biodiversity research and conservation have long understood that the surviving wildlands of the world are not art museums. They are not gardens to be arranged for our delectation. They are not recreation centers or harborers of natural resources or sanatoriums or on undeveloped sites of business opportunities—of any kind. The wildlands and the bulk of Earth's biodiversity protected within them are another world from the one humanity is throwing together pell-mell. What do we receive from them? The stabilization of the global environment they provide, and their very existence, are the gifts they give to us. We are their stewards, not their owners.
>
> — E. O. Wilson, *Half-Earth*, 2016

ACTION TOPIC (IV-5) Create Semi-Wild Patches »

Why This Matters

Some people like their gardens neatly trimmed, while others prefer theirs more loose and naturalistic. We all have preferences, and that's understandable. Here we suggest that every piece of property—from tiny balcony to vast estate and everything in between—probably contains some places that could support at least a bit of un-manicured nature.

The more we can allow organisms in the landscape to interact free of our human intervention, the more nature itself will figure out which things—plants as well as other creatures—are best able to survive and thrive in that vicinity. We might not love everything that nature presents us with, but at least we could give this a chance, just to find out. The idea here is to let ourselves be surprised, in service to the possibility that nature knows best.

If we're talking about a little urban spot with no soil usable for growing things, we have to provide a place for growing things. And then within these

Credit: SReed

Figure IV-17: White wood asters carpet this dry shady bank in late summer. A little later, calico asters and woodland goldenrods will appear. Here, the landowner's only job is to remove unwanted saplings, then sit back and watch the ever-shifting display.

places, we often want things to be just so. This is reasonable. But we can still choose some portion of the total available area where we can stand back a bit, watch what happens, and see if nature fills in the open niche in a way we can tolerate, or even enjoy!

Beyond the city, most suburban landscapes and all rural ones hold the potential for creating a semi-wild haven, corner, nook, or refuge. This might be the farthest place from the house, a shady corner where lawn or flowers won't thrive, that wet spot you keep thinking of turning into a pond, the edges of a utility area, behind the garage, along a narrow strip between two properties. Any unused places can serve as the perfect place to let nature have more of a hand in what happens.

If some plants are allowed to go to seed and produce new generations, the new individuals might contain genetic material that could enable a whole species to adapt in unexpected ways. Natural selection and evolution take place right in your own backyard! Species that need to migrate from elsewhere to escape uninhabitable conditions might find just the haven they need in your landscape—to either support them in a new home or help them on their way farther along.

Actions

▶ **Leave some areas alone.** This is the easiest method; just stop doing much of anything in one particular area. This can be an acceptable option in some places, especially in large natural areas, but it's not the choice that this book recommends, as it could lead to two kinds of undesirable, long-term effects:

1. *An unfettered invasion of non-native invasive plants*, which, if left unchecked for many years, could be-

> Suburbia already has more holes than a slice of imported Swiss, and the routes along solid ground are becoming more and more difficult for animals to negotiate. With animals' fear of exposure and fondness for paths in mind, we can begin to envision basic changes in our landscape. If ordinary garden design begins with the blank space of the lawn which is then cut here and there to create beds of taller plants, we can aim for the obverse: a tall growth of grass, shrubs, and groves cut by mowed or mulched paths that occasionally open into clearings.
>
> —Sara Stein,
> *Noah's Garden*, 1993

come impossible to control when someone finally notices their harmful effect. (See more discussion of invasive plants in IV-4 on page 117 and VII-3 on page 210.)

2. The unwanted predominance of species that *do belong* in that ecosystem but, due to the unnatural conditions, are *not kept in balance*. For example, many small saplings might sprout, either from cut stems of trees that were removed or from an unusual spread of seeds dispersed the previous year, or other plants could emerge from seeds that have been lying dormant in the soil but are released by new conditions.

A better way is to **leave some areas(s) alone and monitor what happens,** and then remove undesirable plants. Some plants, such as annual weeds and aggressive, unwanted perennials will best be removed periodically throughout the growing season. Trees and shrubs might take two, three, or several years

Figure IV-18: Mountain laurels, low-bush blueberry, bush honeysuckle, and hay-scented ferns form a mosaic community full of niches. They also hold the soil tight on this steep reconstructed bank above a trench drain, which diverts runoff and subsurface water away from a house foundation.

Credit: SReed

to reveal their ultimate pattern of growth. We don't recommend removing all woody seedlings and saplings when they first appear, but rather suggest waiting to see how nature itself will respond.

> A third way is not to just stand back but to actually **create conditions that will encourage new things** to germinate and grow. For example, in temperate regions, you might expand a woodland edge by spreading a thick layer of leaves over the intended semi-wild area. After waiting a year or two for essential soil microorganisms to develop, you would then plant in this new soil some species that can naturally extend the forest. Or it could mean trying to recreate growing conditions typical of your own locale, where nature provides and supports vegetation you might like to grow on your property. You could also start by planting a selection of native plants to give the area more definition and a bit of a head start on becoming wild. This option is even more appealing if your property abuts a park or conservation area, allowing your wild space to serve as a transition area between the managed landscape and the larger environment.

The naturally evolved associations of native plants within a particular region can provide both information and inspiration for the design of gardens and landscapes that are ecologically sound and aesthetically satisfying. Furthermore, by utilizing patterns and processes that are intrinsic to naturally evolved landscapes, we can create designed and managed landscapes that are clearly "of the place" and that approach the ideal of sustainability. By conscientiously following this model, we can also protect biological diversity in the human-dominated landscape.

—Darrel Morrison, 1996 preface to reprint of *American Plants for American Gardens* by Edith Roberts and Elsa Rehmann, 1929

❯ **It's also a good idea to link** many semi-wild and wild pieces of nature together to create larger patches and corridors, which will help plants and animals migrate through the landscape to find new habitat if their own home grounds either get eliminated by development or become inhospitable. Read more about habitat corridors in VIII-6.

Last Thoughts

Bear in mind that allowing a portion of your property to be semi-wild is not the same thing as doing nothing. Rather, it means letting natural processes have a little bit more influence over how things go. Somewhat similar to raising children (but easier), it means keeping an eye on things, managing with a light touch, and stepping in when necessary to steer things in the direction you want them to go. And it means enjoying unexpected qualities and characteristics—which you yourself didn't create—that then show up.

V

SOIL

Overview

Soil is an amazing substance. One single handful can contain more organisms than there are humans on the planet. And all of those miniscule organisms work together to form the basis for all life on Earth.

Yet most of us know very little about the soil beneath our feet, and in Western civilizations we have treated soil carelessly for a long time. Thankfully, this is changing rapidly, as we come to appreciate the complex and valuable role that soils play, both in the way plants grow in it and in soil's capacity to absorb excess CO_2 from the atmosphere.

Soil sequesters about four times more carbon than forests and all other vegetation. In fact, after oceans, soil is the second largest CO_2-absorbing medium in the world. *Note*: This figure does not include the amount of carbon stored in rock—by far the largest carbon reservoir on earth—because rock is not absorbing CO_2 in timescales relevant to our discussion. (See "A Primer on Climate Change" on page 6 for more about carbon reservoirs.)

We know that turning/tilling soils releases carbon dioxide into the atmosphere. Some the world's cultivated soils have lost as much as 50% of their original store of carbon.[1] This released carbon dioxide contributes to global warming and acidification of the oceans.

We also know that damaged (i.e., overworked and/or compacted) soil supports fewer microorganisms than does healthy soil, and so ultimately reduces the amount of carbon that is both produced and stored there (see "A Primer on Soil Carbon and Ecosystems" in this section for an explanation of how this works). And soils that are low in carbon tend to have low fertility, which reduces their ability to support small-scale landscapes, vegetation, and crops.

The good news is we know how to put carbon back into the soil, where it belongs. There's a new agricultural movement underway to employ *regenerative practices* that increase carbon storage in soils. But here's the thing: this movement doesn't have to be limited to farming! We can adopt similar principles and actions in domestic and commercial landscapes. This is a good idea for (at least) three reasons:

1. Even though residential and commercial landscapes might not occupy as much land as farmland across the continent, still there are millions of us who can work on this. And there is power in numbers.

2. Unlike agriculture, which must extract matter from the soil to satisfy a financial bottom line, we non-farmers are free to experiment and devote as much or little of our domestic landscapes as we wish, to actions that will increase carbon in the soil.

3. Because most of our non-farm landscapes are located near cities (where atmospheric carbon levels tend to be highest), this is where carbon-sequestering efforts can make a significant difference.

Producing food, growing industrial materials, and sequestering carbon is not enough for a 21st-century farmer. Agriculture must also *adapt* to a changing climate.... Although carbon farming practices aren't necessarily, by definition, adaptive, in practice almost all of them are. This is a great co-benefit of carbon farming: not only do they, by definition, help mitigate climate change, but they also help ecosystems and communities adapt to it.

— Eric Toensmeier,
The Carbon Farming Solution, 2016

In This Section

The following five *Action Topics* explain how to take care of our soils if we want to promote healthier landscapes, support abundant wildlife, and sequester more carbon:

- V-1: Maximize Carbon Storage in Soil
- V-2: Minimize Soil Disturbance
- V-3: Avoid Compacting the Soil
- V-4: Build Compost
- V-5: Cover Bare Soil

More about Soil Microbes and Macrobes

Whenever a seed germinates in healthy soil, the microbial community is activated as a result of the seed's secreting chemical signals into the soil. Sometimes the microbes are bacteria that form root nodules on legumes, which help them fix nitrogen, allowing the legume to grow well in poor soil. But more often it is fungi that form symbiotic relationships with the roots; these fungi are called *mycorrhizae*.

Mycorrhizae relationships work because the fungi help increase the roots' water and nutrient absorption, while the roots provide the fungus with carbohydrates, formed from photosynthesis, that the fungus uses for energy. This specialized relationship sometimes offers the host plant increased protection against certain pathogens. Approximately 90% of all vascular land plants live in some association with fungi.

Macrobes are soil organisms that we can see, such as earthworms, centipedes, insect larvae (grubs), beetles, ants, salamanders, toads, and more, so we probably give them more attention than the microbes. Macrobes are also carbon-based organisms that consume plant matter (both dead and alive). The presence of all these critters means that the soil is being aerated as they rearrange the soil for nests and movement. In the wake of macrobes, microbes make a living from their waste, and the resulting naturally aerated soil provides the perfect medium for plant roots.

Our task is to store more and more of the world's carbon not in the sky (where it disrupts climate) and not in oceans (which it causes to become acidified and damaged) but in soils and vegetation, where that CO_2 is the key to beneficial results, including higher crop yields and more effective water storage. More carbon in the sky is a threat. Carbon in soils and forests is an asset.

—Michael Bloomberg and Carl Pope, *Climate of Hope*, 2017

A PRIMER ON...

Soil Carbon and Ecosystems

Carbon-rich soils tend to support high levels of biodiversity, both above ground and below. Their abundant soil organisms help make nutrients available to plants, build soil aggregates, and increase soil porosity. These soils act like a sponge, soaking up water during rains and flooding, and then releasing it slowly over time. All of these qualities make carbon-rich soils more productive and a vital part of our efforts to curb climate change.

How do soil ecosystems sequester carbon? This elegant process includes the following basic steps:

- As plants grow, they put up to 80% of their energy into producing root systems.
- Plants' roots exude carbohydrates, sugars, and proteins (metabolites), perhaps as a way to grow populations of microbes beneficial to the plant (see below).
- Microbes, including bacteria and fungi, consume these *root exudates* as food.
- Since the soil microbes don't have internal digestive systems, they break down organic matter (i.e., bits of leaves, dead roots, other dead organisms, etc.) externally in the soil. This frees up the nutrients, making them available to be absorbed by the plants' roots.
- The fungi associated with roots (*mycorrhizae*) only associate with new, young roots. They not only help to make nutrients available, but they also help roots absorb water. Some mycorrhizae are specific to only one type of plant, while others support a variety of plant species. As they grow, fungi form filaments, called *hyphae*, which physically create pores in the soil, providing space for yet more aerobic microbial activity.
- In addition to fungi, bacterial microbes create a protective shield around plant roots by creating a sticky substance that also helps to form aggregates in the soil.
- Together, these bacteria and fungi attract predatory organisms like protozoa and nematodes which, as they eat, release nitrogen, phosphorus, sulphur, magnesium, potassium, sodium, iron, zinc, etc., all of which are needed for plant growth.
- And finally, as the result of all these activities in the soil ecosystem, plants grow more roots, which exude more carbohydrates into the soil, which invite more microbes, and so on and so on.

The most important point here is this: when organic matter breaks down, the carbon in its tissues becomes part of the soil humus. This means it gets stored (sequestered) in the soil. If the soil is left undisturbed, most of the stored carbon stays in place. However, if the soil is tilled, turned, or dug, its carbon binds with oxygen in the air to form carbon dioxide, which is emitted as a gas to the atmosphere. (*Note*: Anaerobic decomposition releases small quantities of the even more potent greenhouse gas, methane.) Our goal is maximize that sequestered carbon in the soil, to minimize our CO_2 emissions.

How Many Microorganisms Are We Talking About?

One gram (about ⅕ teaspoon) of healthy soil could contain one hundred million bacteria, one million actinomycetes, and one hundred thousand fungi, whose filaments, if strung together, would measure about 16 feet in length. This same gram of soil could also contain hundreds of nematodes living on the damp surfaces of the soil particles and maybe a few insect eggs or larvae and some earthworm cocoons. The exact proportions of each of these organisms depend on soil conditions such as moisture, aeration, amount of humus, and the local plant community growing in the soil.

Photosynthesis

Green plants have the ability to take in carbon dioxide and water and, combined with energy from the sun, to form sugar with oxygen as a by-product. Plants can then create more complex carbon-based chemicals from the simple sugars and other nutrients to serve their needs, which can be quite complex. In turn, almost all forms of life on the planet including animals, plants, bacteria, fungi, and protists (microscopic and unicellular organisms) gain energy for life from ingesting plant-based sugars directly or indirectly through respiration—the equal and opposite reaction of photosynthesis. It's important to remember that an important "by-product" of photosynthesis is the oxygen that all plants release into the atmosphere.

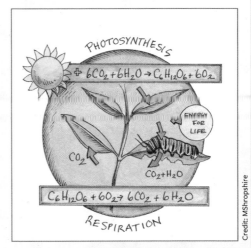

PHOTOSYNTHESIS

$$6CO_2 + 6H_2O \rightarrow C_6H_{12}O_6 + 6O_2$$

ENERGY FOR LIFE

CO_2

$CO_2 + H_2O$

$$C_6H_{12}O_6 + 6O_2 \rightarrow 6CO_2 + 6H_2O$$

RESPIRATION

Credit: MShropshire

Plants create sugars through photosynthesis, using water and carbon dioxide as base materials. Almost all living things on the planet use those sugars directly or indirectly to gain energy for living, and some of the carbon in those sugars (carbohydrates) gets stored in the soil.

ACTION TOPIC (V-1) Maximize Carbon Storage in Soil »»

Why This Matters

In nature, soil is one of the main places that carbon accumulates; it is the result of plant photosynthesis (see above). However, the development of cities and suburbs, along with centuries of farming that tilled and turned soil, have now released large amounts of historically sequestered carbon into the atmosphere. Studies differ about how much this carbon dioxide contributes to global warming, but all agree the amount is substantial. So, it just makes sense to put some of that atmospheric carbon back where it came from: into the ground.

Even though soil is such an important asset for our survival, for the most part, through the ages, humans have treated our valuable soils like dirt. In ancient times, slash and burn was the only way we knew to survive. But it turned out that removing and burning forests greatly reduced the production

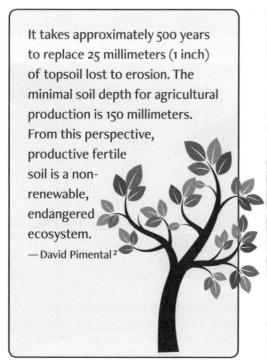

It takes approximately 500 years to replace 25 millimeters (1 inch) of topsoil lost to erosion. The minimal soil depth for agricultural production is 150 millimeters. From this perspective, productive fertile soil is a non-renewable, endangered ecosystem.

—David Pimental[2]

Credit: GStibolt

Figure V-1: This Costa Rican soil layer is thin because the high metabolism of tropical rainforest ecosystems breaks down soil humus very quickly and releases its carbon into the atmosphere. As the climate warms, more soil will become thinner unless we take good care of it. Read on!

of humus. As a result, the soil ecosystems were less able to sequester carbon and became less fertile. This might have been forgiven when humans didn't know any better. Today, though, we do know the consequences of our actions. Instead of toxic lawn care procedures, continued massive deforestation, and unsustainable agricultural practices, we could be doing much better.

The good news is that soil can be rehabilitated. Even in big commercial agriculture and forestry activities, efforts are being made to rejuvenate and enrich the soil by tilling less and using cover crops. Forests are being replanted, and while these efforts are long-term projects, there *are* signs of hope. This book concentrates on doable management actions for smaller property managers such as homeowners, communities, schools, churches, and municipalities. Many of these actions are easy to implement, and, in addition to taking better care of your soil in the abstract sense, they may even save you money and improve your health.

Actions

● **Grow more plants.** When there is a dense cover of diverse plant species, more carbon can be stored in the soil because:

- Roots are a source of carbon in the soil.
- A dense plant population encourages the growth of mycorrhizae and other soil microbes, which increases carbon in soil.
- Cooler, moister soil under the plants in turn supports even more vegetation that increases the carbon storage in the soil.
- Plants add organic matter to the soil with their leaf and litter drop, which improves its structure so it can store more carbon.

● **Avoid using synthetic fertilizers.** Added nitrogen stimulates the soil's decomposers whose activity accelerates the release of carbon held in the soil by plant roots and

Credit: GStibolt

Figure V-2: At the Jacksonville Zoo, this savannah is made up of densely planted Florida natives. Children are encouraged to run through this low maintenance meadow, which is mowed once a year. Although the animals at the zoo might be exotic, the plants are native.

Figure V-3: A dense flower bank in a Midwestern urban space includes black-eyed Susan and bunching grasses, and leaves no room for weeds.

Credit: GStibolt

other buried organic matter. In addition, synthetic fertilizer does not include any humus or substrate as organic sources do, and consequently is readily rinsed out of the soil during rain or irrigation. The resulting loss of organic materials also makes soil more susceptible to compaction and erosion.[3]

Using organic fertilizers will tie the nutrients to humus; thus, they remain available longer and able to help restore the soil's microbes. The result is that the plants and the soils become more resilient and store more carbon. (See "A Primer on Landscape Chemicals" on page 19 for more information on synthetic vs. organic fertilizers, and V-4 on page 144 for information on building and using compost in the landscape.)

⊘ **Avoid using peat moss and peat products.** Peat bogs and peat lands store vast amounts of carbon. There is no sustainable way to harvest peat. This substance takes hundreds of years to form under special anaerobic conditions, and efforts to restore mined peat fields results in more CO_2 being released than sequestered.

Climate-wise landscaping excludes peat products and the use of peat moss as a soil amendment. Yes, peat moss absorbs moisture, but it is extremely acidic and provides virtually no nutrients. A viable substitute might be coconut coir, a by-product of the coconut industry. Absorbent and neutral in its acidity, coir also provides nutrients. The one big drawback of coir is its transportation footprint, because most of it is produced in Indonesia. The most

Credit: SJeed

Figure V-4: (*Above*) This thriving bog in New Brunswick, Canada, sequesters large amounts of carbon in the ground, where soil acidity inhibits the microbial activity that would release carbon dioxide. Such ecosystems also contain a rare assemblage of plants and animals.

(*Below*) When stripped to produce peat moss, all of those organisms disappear, and the stored soil carbon is lost to future breakdown and release. Plus, the resulting denuded soil may take hundreds of years to regenerate. If possible, avoid using peat moss products.

Peat lands store as much as 500 billion metric tons of carbon— or twice as much as is incorporated into all the trees in all the world's forests—roughly 1,450 metric tons of carbon per hectare. And the United Nations Environment Program estimates that reducing global deforestation, especially that occurring on top of peat lands, could restore some 50 billion metric tons of CO_2, or nearly two years of global emissions. Although peat lands do emit methane—a potent greenhouse gas—this is more than outweighed, in terms of the overall balance of greenhouse gases in the atmosphere, by the carbon dioxide they sequester.

— David Biello [4]

climate-friendly soil amendment is leaves rotted for at least one year, which are effective in soil mixes, compost, and mulch.

All of the *Actions* in the rest of this section—which address reducing soil disruption and compaction, and building soil structure, fertility, and microbial activity—contribute either directly or indirectly to building soil carbon.

Last Thoughts

Good landscape stewards need to be good stewards of the soil. The health of soil ecosystems, with all their microbes and macrobes, is key to the vitality of landscape plants. In other words, if you take care of the soil, it will take care of your plants. In addition, a healthy soil sequesters a large volume of carbon in the form of humus and in all the critters that depend on the sugars created through photosynthesis.

ACTION TOPIC (**V-2**) Minimize Soil Disturbance »

Why This Matters

Soil is the most important terrestrial carbon sink, but it emits CO_2 on a continuous basis as its inhabitants—plant roots, microbes, and macrobes—respire. This release of CO_2 is exacerbated when soil is disturbed and soil aggregates are broken. Further, turning soil dries it out more quickly, which releases its stored moisture as vapor into the air.

Higher CO_2 emissions in disturbed soil are due to many factors, including:

- Increased metabolic activity of microbes in the highly aerated environment.
- New opportunities for microbes and macrobes to work on the decomposition of plants and animals damaged by the disturbance.
- The quick germination of previously dormant seeds (mostly opportunistic weeds), which triggers intense microbial activity in the newly formed rhizospheres.
- The physical act of breaking open previously sealed pores where gases—CO_2, nitrogen, possibly methane, and others—were stored.

When we minimize the disruption of the soil ecosystem, the soil can continue to sequester CO_2 and other greenhouse gases. Also, plants do better if the soil fungi and all their underground support systems remain intact. In turn, healthier plants maintain or increase their carbon uptake and can better support wildlife.

Actions

Reduce (or stop) decorative seasonal plantings. Short-term plantings—such as begonias, impatiens, pansies, and other seasonal plants—that are installed in full bloom and then replaced a month or two later cause repeated soil disruptions. This happens each time one set of tired-out plants are replaced with new, blooming ones. The soil's ecosystem doesn't have time to recover, and high amounts of CO_2 are emitted continuously.

Replace these temporary plantings with long-lived perennials and woody plants so the soil can reach equilibrium. See Section VII for suggestions about long-term solutions.

Municipalities, communities, and commercial properties can save money and reduce greenhouse gas emissions by redesigning their managed properties to include mostly native, more permanent plantings and by rewriting their maintenance contracts to eliminate seasonal plantings.

Credit: GStibolt

Figure V-5: Seasonal plantings of ornamental annuals might be replaced two or three times a year in some regions, for continued display. This disturbs the soil and releases more CO_2 than would perennial plantings, which don't need to be replaced.

❯ **Buy small plants.** If you are buying plants instead of growing from seed (the most climate-wise choice), there are several good reasons to purchase young plants in small pots:

- Installing small specimens disturbs less soil than planting larger ones.
- Smaller, younger plants need less supplemental irrigation to become established.
- Smaller, younger plants become established much faster than larger specimens and so begin to supply ecosystem services more quickly.
- Their survival rate is much higher than for larger specimens, especially for those trees and shrubs that have been kept in containers for several years.
- The ecological footprint of growing and transporting smaller plants is correspondingly smaller.
- You save time and money.

Figure V-6: If you must pull weeds, use the two-finger approach, which reduces soil disturbance because your fingers keep most of the soil in place. If you know a weed is an annual, cut it off at its base instead.

Credit: MShropshire

❯ **Reduce the need for weeding, and weed more gently.** Every time you pull up a weed, the soil is disrupted, and every time the soil is disrupted, CO_2 is released and more weeds are likely to grow. So use mulch, get rid of weeds before they go to seed, and space herbaceous plants closely, so there's more shade and less room for weeds.

❯ **Minimize soil disruption when growing edibles.** Use raised beds and other intensive methods of gardening to grow crops at the highest density; this will allow you to use, irrigate, and disturb the least amount of soil. Raised beds make it easier to tend to crops without stepping on the planting surface. A home garden should not look like a miniature farm with long rows widely spaced on a large piece of land. And no roto-tilling or other deep tilling should be done once the beds are built and put to use. See Section IX for more ideas on adding sustainability to our edibles.

▶ **Minimize importing topsoil.** Soil moved from somewhere else produces greenhouse gas emissions due to its excavation, transport, and the process of grading it once on site. A better approach is to slowly build up the soil's nutrients and microbe/macrobe populations with top-dressings of locally made compost. Also, by adding plants to the landscape that don't require high nutrient levels in the soil, there is less reason to import it. Learn to love—and build up—the soil you have.

Last Thoughts

Most landscapes can be managed with a light touch by mimicking natural processes, and by adding compost and mulch to the top of the soil. This material will soon become part of the duff (the loose litter of leaves and twigs that lay on top of soil.) and will be incorporated into the soil's ecosystem by the action of abundant microorganisms. All of these actions bolster soil's ability to support the greatest amount and diversity of plant and animal life, both above and below ground.

Credit: GStibolt

Figure V-7: Wide-row planting provides trenches between the rows for better drainage and for more room between crops—for larger plants, the trenches would be wider.

The creation of the fertile soil, the enhancement of fertility in existing soils, and the re-creation of fertility in worn out soils, are the only means by which humankind can immediately, right now, today, put the stops on our runaway global warming. Increasing soil fertility necessitates absorbing and sequestering the damaging carbon dioxide gas from the atmosphere into the ground. This is the only way we can obtain some time to allow us to build our new, non-fossil-fueled energy systems; the system that will allow us to run our civilization on energy sources that don't change the nature of the very air we breathe.

— Allan Yeomans, *Priority One: Together We Can Beat Global Warming*, 2007

ACTION TOPIC (V-3) Avoid Compacting the Soil »

Why This Matters

Healthy soils normally contain more than 40% *pore space*: air holes of various sizes, throughout the top layer. The larger spaces promote drainage, and the smaller pores help store water. Together, they enable air and water to penetrate the soil and promote good drainage, both of which are necessary for soil organisms to live, and for roots to grow.

Repeated compressions due to machinery, vehicle, or foot traffic compact the soil and reduce the soil's porosity. Even pounding rain will compact unprotected soil. This can limit the amount of oxygen available for respiration by soil microbes and macrobes, which will reduce the biological activity of the whole soil ecosystem, further impeding the ability of plant roots to grow well and deeply. In addition, compacted soils are susceptible to both flooding and drying out, since water runs off rather than infiltrating.

Figure V-8: Parking cars on lawn can compress the soil, but if the parking is infrequent, a diverse lawn like this one will absorb rainfall and serve the local ecosystem between the parking events.

Credit: GStibolt

Actions

▶ **Minimize damage to soil by directing traffic in the landscape.** Create obvious pathways and designated areas for walking and other outdoor activities for people and pets. Cover the soil of these paths and outdoor rooms with a thick layer of persistent mulch and use stepping stones to spread out the pressure. If the soil is wet on a regular basis, build raised walkways and decks to keep foot traffic off the soil altogether.

When designing traffic areas, pay attention to areas where tree roots roam. Trees have wide-ranging root systems, which is good for drought tolerance and wind resistance. Protect these roots by keeping the traffic away from them and by planting understory shrubs and plants so the area is not so tempting to walk upon. (See IV-2 on page 109 for more ideas about grouping plants.)

⊙ **When soil is wet, stay off.** Soil is more easily compacted when it's saturated, so limit gardening and landscaping activities to times when the soil is relatively dry. This is especially true for silty and clayey soils; their tiny particles tend to quickly migrate into all available pores when rain or floodwaters percolate through.

⊙ **Vary mowing patterns,** whether you use a riding or push mower. This way, the soil under the lawn has a chance to recover. Also, as discussed in Section I, mow less often; this is better for the plants in the lawn *and* for the soil underneath.

⊙ **Prepare the landscape in advance of projects that could compact the soil.** If you're planning a big project where there will be lots of extra foot traffic and extra machinery, cover the soil with a thick layer—at least six inches—of wood chips or other bulky, organic mulch. You may want to lay sheets of scrap plywood in areas where most of the machinery and vehicles will be working to spread out the pressure. (See II-1 on page 52 for more about protecting tree roots before

Credit: GStibolt

Figure V-9: Standing rainwater in this Florida swale compresses the soil, and later, when lawn crews try to mow here, the soil is compressed even more. A more climate-wise approach would be to create a low-growing rain garden of native rushes, sedges, water-tolerant bunching grasses, and water-tolerant forbs such as hibiscus.

Credit: SReed

Figure V-10: Mowing lawn in the same way repeatedly will compress soil and impact its health. Experienced lawn crews make a point of changing the mowing pattern every time. (But even better is to have a smaller lawn than this one.)

and during construction.) When the project is completed, remove all but two inches of the mulch, as you get ready for your next phase of landscaping in that area.

> ⟫ **Remedy compacted soil.** It's much easier to prevent compressing the soil than it is to loosen it afterward, but these actions can help:
>   • Add a thin layer of completed compost—one or two inches—over a whole area of compacted soil and then lightly irrigate. *Note*: Only add compost to soil when no rain is expected for at least two days. You don't want this fresh compost to end up in nearby waterways if more water falls than can be absorbed. (See V-4 on page 144 for information on compost.)
>   • Create holes in the soil. There are two types of aeration machines—one type removes small cores of soil, while the other pokes holes in the soil, allowing water and air to penetrate into the soil. If you will be adding compost to the area, drill the holes first. These machines are appropriate only for lawn or meadow areas, not for areas where there are a lot of tree roots.
>   • Don't till the soil. Tilling might be appropriate in large agricultural operations, but try to avoid it in home or commercial landscapes. Tillage releases CO_2 and damages the soil structure and the microbes that are in place.

Last Thoughts

Soil and soil organisms provide the basis for all that is alive on the planet, from the tiniest microscopic flagellates and bacteria to the most majestic sequoias and redwoods. Designing and caring for our landscapes specifically to protect soil from compaction—to keep soil alive and vibrant—reminds us of the importance of this reality.

ACTION TOPIC (**V-4**) **Build Compost** ⟫⟫

Why This Matters

When organic materials are recycled into the soil, the soil's ecosystem will be improved and will be better able to sequester CO_2 and other greenhouse gases.

Further, if soil enrichment is accomplished with local, free materials and without the aid of synthetic fertilizers, energy, carbon emissions, and costs from both the manufacture and the transportation of these products will be saved.

Traditional landscaping practices in temperate regions often include raking up grass clippings and leaves in the fall, bagging them (often in plastic), and then putting them on the curb for collection. In a best-case scenario, the municipality might add those leaves to a huge, hot compost pile that is then used in municipal projects. In many cases, though, the leaves go to the landfill and are buried under garbage where they will rot anaerobically, which generates methane, one of the most potent greenhouse gas.

Many modern households have garbage disposals where most of the kitchen scraps are added to the septic tank or sewer system. You can reduce the volume of your waste stream by incorporating your kitchen scraps into your landscape through composting in a general compost pile, or by burying them directly in edible gardens between planting areas.

Also, as discussed near the beginning of this book in "A Primer on Landscape Chemicals," the use of synthetic fertilizers causes more pollution of groundwater and nearby waterways, and it alters the structure of soil, which reduces its ability to store carbon. Using compost in the landscape is an ideal antidote to all of these problems.

Actions

⊙ **Save leaves and other landscape trimmings.** Leave the leaves where they fall in groves or wilder areas of your landscape because here they provide the perfect mulch for the trees and shrubs that have shed them. (See more about planting in groups and groves in IV-2 on page 109.) Fallen leaves also contain the unseen eggs and larvae of many overwintering pollinators and other insects that are easily harmed by raking, so keep raking to a minimum. Rake leaves that fall on lawns, patios, sidewalks, driveways, and other hardscape features and save them in a pile for later use as mulch or to build compost. If you need a greater volume of compost than can be created from your own leaves, ask your neighbors for theirs. Some municipalities make collected yard waste available to the public—before or after composting. (See below.) Be careful, though, about using yard waste from unknown sources; you want to avoid importing unwanted weed seeds, invasive plants, and chemical residues.

❯ **Save and use your kitchen scraps.** Cooking meals from scratch is good for the planet for a number of reasons, as discussed in Section IX, but meal preparation will generate peels, seeds, cores, stems, and other trimmings. Instead of stuffing them into the garbage disposal, save them in a bucket or container and then add them to a compost pile or bury them directly in edible gardens between your crops.

Figure V-11: Trench composting can take many forms. Pictured here is burying kitchen scraps in the bottom of a trench between wide rows in a raised bed. The trenches provide good drainage between the planted surfaces, while adding this extra organic material enriches the garden as a whole.

Credit: GStibolt

❯ **Build a good compost pile.** While there are many methods of setting up a compost system, here are some general composting guidelines:

- General rule of thumb is to use equal amounts of green and brown materials in alternating layers when constructing a new pile. If you reach the right balance, the pile will become hot within a day or two as the microbes start working.
  - *Brown materials* are mostly dry and are high in carbon, such as dead leaves, wood chips, pine needles, straw, and shredded paper.

Vermiculture

Vermiculture is composting with special worms, which are available online or in specialty garden shops. Usually people use this method for composting their kitchen scraps. Use shredded and dampened black ink-only newspaper as brown material and the kitchen scraps as the green material, maybe with a little soil and composted manure to balance things out. To harvest the compost, lure the worms to new food (in a separate section of the worm bin), and then clean up their leavings. Vermiculture requires ongoing attention to make sure the worms don't starve or dry out. The resulting dark, crumbly material is the black gold of composting. However, for best results, it should not be used alone as a growing medium but instead mixed with at least some amount of soil or compost from a compost pile. Do not release the worms into the landscape.

- *Green materials* are softer, have more moisture, and are higher in nitrogen, such as freshly pulled weeds, grass clippings, coffee grounds, kitchen scraps, and manure.
- The more often you turn the pile, the faster it will become fully composted. (Turning is rearranging the pile so the stuff that was on top is on the bottom, and the stuff that was on the outside is moved to the inside.)
- If your compost smells like ammonia, it has too much green material. If the compost smells sour or more like rotten eggs, this means that *anaerobic decomposers* are working and greenhouse gases are being emitted. To solve either of these odor problems, add some brown material and turn the pile to introduce air; leave the pile uncovered until it becomes just moist to the touch. A good compost pile should not smell bad.

Test the Soil

Knowing the nature of the soil will help inform the choice of which plants will have the best chances of long-term success. Soil test kits are available online, at the local extension office, or in local garden shops. For general landscape uses, soil should be tested for acidity and macro- and micro-nutrients; in urban areas, you should also test for heavy metals or other toxins. Sample the soil in several places to obtain a good average of the soil.

Soil acidity is measured on the pH (*potential of Hydrogen*) scale, with a pH of 1 being most acidic and a pH of 14 being most alkaline. Most plants, including crops, grow well in a slightly acidic soil, meaning a pH between 6 and 7, but some plants are adapted to highly acidic soils and other plants thrive in alkaline soils. The minerals and/or the humus in soil determine its acidity. This can't easily be changed for the long run, so it's best to live with the existing soil and use appropriate plants.

Soil texture depends on relative proportions of sand, silt, and clay particles. These proportions are also not easy to change. Although it might seem logical, it doesn't work to add sand to a clay soil to make it more porous; the end result may be a hard, cement-like material. The best way to improve soil structure, whether it is sandy or clayey, is to add locally made compost. In the natural, native, or mostly native sections of your landscape, one good option is to let the soil reach its own balance by letting leaves stay on the ground where they fall, to gradually break down and return their nutrients to the soil.

• For native or near-native landscaping, don't use manures in the compost. For edible gardens where you need more fertile compost, you may add manure from rabbits, chickens, horses, cows, or other herbivores, but don't use pet or human feces. They can introduce harmful bacteria to the soil. Human urine is okay, though.

• Do add kitchen scraps, but not meat, oils, or dairy products. This keeps the odors down, and it also discourages raccoons, opossums, crows, and other scavengers from raiding your pile.

• Don't add lime to your compost pile even if your original materials are acidic. The decomposers will neutralize most of the acidity, and lime might kill many of the decomposers.

❯ **Use compost in the landscape.** Keep in mind that compost is not used in the same manner as an artificial fertilizer. Instead, it is applied more generously, as a way to improve soil texture and structure along with providing nutrients. Apply compost as a top-dressing to the landscape; to keep soil disturbance to a minimum, do not dig it in. The compost will be absorbed into the soil itself as its soil organisms work. The organic materials in the compost provide the food for the soil's ecosystem. There are many ways to use compost. You can:

• *Work it into vegetable beds.* After initial clearing of weeds or digging in the green manure, apply three inches in the planting areas and work it gently into soil.

• *Use it to condition soil in and around perennials and shrubs.* If the bed is covered with a layer of persistent mulch, gently scrape the mulch away, apply a ring of compost around each plant out to the drip line or more (but not touching the stem or trunk of the plant), and then replace the mulch.

• *Fertilize trees* by laying a wide ring of compost one to two inches thick at and outside of the planting hole and

Figure V-12: Adding compost to a raised bed prepares it for the next set of crops. Since the bed is raised, crops can be grown close together, and underlying soil is not disturbed. Growers walk on mulch-covered pathways between the beds, preventing compaction.

Credit: GStibolt

Credit: GStibolt

Figure V-13:
Compost bins can
be made in many
ways, but using
recycled pallets is a
particularly climate-
wise approach, as
long as the wood has
not been pressure-
treated (see X-2 for
more about wood).

eventually out to the drip line. Don't dig the compost in; either cover the
compost with a mulch or leave it as is.

- Compost can be *used as a mulch by itself*, even if it's unfinished and you can
  still identify some of the original materials. It is moderately effective as a
  weed suppressant, but it doesn't do much to retain moisture or moderate
  soil temperature fluctuations.

- *Use it for container gardens*, either as a top-dressing or as a planting medium.
  It's particularly important for long-term container plants to have a living
  soil. Potting soil or soil-less mixtures tend to deteriorate; compost adds
  structure and microbes to the soil, which will keep the soil working and
  reduce the frequency of re-potting.

- Apply a two- or three-inch layer of compost on the soil before planting new
  lawn to provide a good healthy soil for the grasses and other plants.

Last Thoughts

Adding compost to the soils in your landscape adds some nutrients and, more
important, it keeps their ecosystems alive so that the whole complex web of
inhabitants will work together to support maximum plant growth and sequester as much soil carbon as possible.

ACTION TOPIC **V-5** **Cover Bare Soil**

Why This Matters

Keeping bare soil covered is the best way to prevent erosion, preserve soil moisture, and protect or enhance soil ecosystems, which in turn promote healthy root growth. This contributes to the development of soil carbon, which in turn helps mitigate climate change. Another reason to cover bare soil in gardens is to prevent weeds.

One of the best ways to cover soil is to grow an abundance of plants that spread and arch over the ground, leaving little unshaded space between. Increasingly landscape designers are creating densely planted gardens which include certain species that establish quickly, along with others that might start slow but will later outcompete their early-starting companions.

More typical in newly planted gardens is to position plants according to their expected mature size and cover the ground between with some form of mulch.

Mulching works to suppress weeds in two ways:

- It creates a physical and a light barrier, which helps weed seeds stay dormant. Of course, this does not prevent weed seeds from falling on top of the mulch where they can germinate—whether or not the mulch is organic.
- Organic mulch materials also work chemically. As they decompose, their microbes absorb nitrogen from the soil to live; this temporary nitrogen depletion suppresses weed seed germination and may even kill some weed seeds.

Actions

▶ Remove aggressive weeds before mulching. Mulch may be used to keep down weeds in an established bed or to start a new bed or other non-lawn area. First, remove highly aggressive weeds with deep rhizomes from the area because mulch will not impede their growth. Top with a thick layer (four or five inches) of mulch such as leaves, pine needles, or tree trimmings.

▶ Mulch should not be a main feature of the landscape. Removing lawns is a great idea and mulch is certainly recommended between plants to reduce weeds, keep in moisture, and to improve the soil. But a mostly mulch landscape is not a reasonable answer because it's not an inviting ecosystem for pollinators and other wildlife. A mostly mulch landscape ultimately invites weeds in all but the most arid climates because many annuals can find a foothold in the least bit of soil between mulch particles. The best solution is to grow more plants to shade the soil and leave little space for weeds. (See Section VII and IV-3 on page 113 for further discussion.)

Credit: GStibolt

Figure V-14: This foundation bed is densely planted with grasses and yuccas. There is mulch here, but it's not the focus of the landscape—the plants are.

🞂 **Prevent mulch from being in contact with woody plant stems and trunks.** It's not uncommon to see a foot or more of mulch leaning against tree trunk (this is often called *volcano mulching*). This is harmful to trees in several ways:

- Water may be shed from the mulch away from the tree, because of the steep angle.
- Rodents and other animals can burrow and nest in mounded mulch and chew the bark.
- Fungi harbored in the mulch could rot out the base of the tree, making it vulnerable to insect invasions and disease.
- Some trees and shrubs will create new roots within a thick mulch layer. Plants' roots should be in the soil for the best stability and the best drought resistance; in droughts, the mulch will dry out first.

🞂 **Avoid using newspaper or cardboard** under mulch, because these "sheet mulches" can induce anaerobic conditions if used on wet, poorly drained soils. When wet, the layers of

Credit: SReed

Figure V-15: Mulch "volcanoes" are an unfortunate landscaping tradition: organic material heaped against a trunk not only does no good for the tree roots, (which extend far beyond the tiny central mound), but it actually hurts the bark, by causing rot, which invites disease and pests.

paper become compacted, creating an impermeable barrier to water and gas exchange; when dry, they become *hydrophobic*, causing rainfall or irrigation water to sheet away rather than percolate through.[5]

Credit: GStibolt

Figure V-16: Weed barrier cloth, which is often recommended to prevent weeds, often allows weeds to grow in the mulch and soil above. Roots also can grow through the cloth itself, potentially hurting plants and making later removal of the fabric (or plants) difficult.

◉ **Avoid using weed barrier cloth or geotextiles** (woven or non-woven) under mulch because roots of both the desired plants and the weeds grow into and become enmeshed in the cloth, so the cloth does little to discourage the weeds. The fabric can also girdle tree roots and trunks as they grow. Also, animals that tunnel in the soil will find an edge or a seam and push it out of their way. In addition, it can be difficult to remove this fabric, and, especially if there are woody plants in the area, significant root damage can occur when the sheeting is removed. Finally, these fabrics are non-recyclable and made from fossil fuels, so they have a large climate footprint. A better approach is to minimize weeds through dense plantings that shade the ground.

Last Thoughts

Protecting soil, especially with smart mulching practices, plays a big role in any climate-wise landscape—on many levels. Most important for our discussion, though, is that these practices protect and enrich the soil so it can store more carbon and better support the plants that we need for cooling the air, the ground, and our homes and buildings.

Additional Resources

- The National Pesticide Information Center: npic.orst.edu
- *The Soil Biology Primer,* by Elaine R. Inghan, Oregon State University extension.illinois.edu
- *Teaming with Microbes: The Organic Gardener's Guide to the Soil Food Web (Revised Edition).* Jeff Lowenfels and Wayne Lewis. Timber Press, 2010.

Types of Mulch

Organic mulches hold in moisture, prevent the soil surface from crusting, help prevent erosion on level grades or gentle slopes, protect trees and other plants from mowing injuries, and eventually add compost to the soil. *Note*: Bagged mulch, *no matter what type*, uses energy in its packaging and delivery. Organic mulch options include:

- *Arborist's wood chips*: a mix of shredded wood and green leaves. This is

Credit: GStibolt

Figure V-17: Arborists are usually happy to leave wood chips and trimmings in the landscape or neighborhood where they are working. This is climate-wise mulch, because the trimmer doesn't make the trip to the dump, you don't make the trip to a garden center to purchase chips, and no energy is spent on packaging.

one of the more sustainable mulching solutions because arborists save gas by dumping their loads in the neighborhood; we save gas and money by not paying someone to deliver mulch; and landfills are not used.

- *Compost*: Unfinished compost (which has some of the original ingredients still visible) provides good, soil-enriching mulch. It suppresses weeds to some degree, but it doesn't moderate temperature or moisture.

- *Leaves or shredded leaves*: It's best to let fallen leaves remain because they supply just the right nutrients, augmenting the soil ecosystem. Fallen leaves also contain an ecosystem of their own, often supporting many overwintering insects, both in egg and larvae form. Large or leathery leaves should be shredded; they don't last as long this way, but they do stay in place better than whole leaves, and shredding prevents matting.

- *Pine needles*: A long lasting and good-looking mulch; their waxy or resinous coating makes them drought resistant and also helps the needles last longer than most other organic mulches. Pine needles will only slightly lower soil pH (increase the acidity) because they break down so slowly.

- *Hay or straw*: Hay consists of grasses that have been allowed to go to seed before mowing and gathering. Straw generally does not include the tops or seed heads, but just the grass stalks, so it should have fewer weed seeds.
- *Pine bark or other bark*: Bark does not absorb as much water as whole-wood products because it contains more waxy materials. Use only a two- or three-inch layer of bark, which will last longer than other organic materials. Don't use bark for mulching rain gardens or bioswales, because it will float away.
- *Shredded cypress wood*: Cypress has long been a popular mulching material, dyed in many colors. The wood resists decomposition, holds its form, and holds moisture well. Historically it was a by-product from lumber mills, but now whole cypress forests are being cut down to satisfy customer demand. So while this material makes good mulch, using it is no longer sustainable or climate-wise.
- *Sawdust*: A by-product from lumber mills, the small particles pack together and often cause anaerobic conditions if held in a pile, which releases greenhouse gases. It reduces nitrogen from surrounding soils more

than other mulches, so it's a good weed suppressant and path mulch, but should be used with care around sensitive plants.
- *Cocoa shells*: This is a by-product from chocolate-processing. It is effective as mulch and has some allelopathic properties to reduce seed germination. If available from a local source, cocoa shells are a climate-wise choice. (*Note*: Don't use cocoa shells if you have a dog.)

Inorganic mulches protect soil from erosion and offer some protection against moisture loss and weed growth, but they don't add to the carbon in the soil. Nevertheless, inorganic mulches may be the preferred mulch for arid areas with nutrient-poor soils. Options include:

- *Crushed stone and rock*: Often recommended for succulent gardens, xeriscaping, and fire-wise landscaping. In temperate climates, stone is not much of a weed deterrent. Ongoing maintenance may include leaf blowers and herbicides, both of which make landscapes less climate-wise. Also, rocks retain heat long after the sun has set, and too many rocks can turn a sunny spot into an oven. And after crushed stone is added to the landscape, removing it is difficult.

- *Shredded rubber*: Recycling tires is an admirable goal, but shredded rubber as mulch is not a good idea because it leaches potentially harmful and persistent chemicals into the soil. In addition, the material is highly flammable.

- *Plastic sheeting*: Used as mulch, plastic sheeting is typically stapled to the soil between vegetable rows. This requires a drip irrigation system because the plastic doesn't allow water to penetrate. Plastic is a petroleum-based product, and its manufacture has a large carbon footprint.

PLANNING & DESIGN

Overview

When it comes to the design of landscapes, planning ahead has always been a good idea. Now in this era of global climate change, trying to anticipate and accommodate uncertain and perhaps difficult new conditions is more important than ever. Of course, the desire to create aesthetically pleasing landscapes will always be important. But if we design with forethought about how to make our yards and gardens as resilient, adaptable, functional, and energy-efficient as possible, the results will also be more likely to meet our aesthetic hopes and expectations.

This section of the book is about planning the overall arrangement of the major components of our landscapes, *not* about designing gardens in particular. The *Actions* presented here will help us create landscapes that, in addition to being visually pleasing, will also:

- Save money and lower utility bills.
- Prevent disappointments and future needs for repair and replacement.
- Minimize damage to the site and the cost of later repair.
- Be as easy to construct as possible.
- Help shrink a landscape's carbon footprint.

In This Section

To help achieve these goals, this Section presents the following *Action Topics*:

- VI-1: Design New Home Sites with Climate in Mind
- VI-2: Fit Landscape to Land

- VI-3: Design for Multiple Purposes
- VI-4: Create Energy-Wise Landscapes
- VI-5: Design Flood-Wise Landscapes
- VI-6: Install Buffer Zones
- VI-7: Create Fire-Wise Landscapes
- VI-8: Incorporate Renewable Energy
- VI-9: Design Climate-Wise Driveways and Parking

ACTION TOPIC (VI-1) **Design New Home Sites with Climate in Mind** »

Why This Matters

Whenever we buy or build a new home, we are already thinking about the future. Now, as we look ahead to a changing climate, it makes sense to factor in the likelihood of hotter summers, longer dry periods punctuated by more intense storms and, in some regions, the increased possibility of flooding, drought, fire, and other conditions that could affect both our home and our landscape.

Many books and other resources show how to design buildings for energy-efficiency and reduced heating/cooling costs. The goal—which is a goal of this book as well—is not only to save money, but also to shrink our climate footprint by reducing our own carbon dioxide emissions (either at home or at the power station). We can work toward these same ideals in our landscapes, too. Here, we present guidelines for how to position a new home and plan the whole site to get the greatest benefit from our carefully designed, energy-efficient buildings.

> On no account place buildings in the places which are most beautiful. In fact, do the opposite. Consider the site and its buildings as a single living organism. Leave those areas that are the most precious, beautiful, comfortable, and healthy as they are, and build new structures in those parts of the site which are least pleasant now.
>
> —Christopher Alexander, *A Pattern Language*, 1977

Actions

◐ **Orient the house** so that a wall with many windows faces generally south, to maximize winter warmth. Although we realize this might not be possible on small lots or in subdivisions where zoning or by-laws constrain the positioning of the house, or where trees or other houses might block the southern sun, this is a good idea for the rest of us to keep in mind. *Note*: In warmer climates, maybe a windowed wall would face east or northeast to avoid the hot sun during the summer.

- **Place the house midway on the slope** if a site contains a slope. Too near the bottom may lead to surface runoff accumulating behind the house (or in the basement). Too near the top could expose the building to cold winter wind. One exception to this rule is if the slope faces north; in that case, a house nearer the top will get maximum winter sunlight.

- **Capture natural ventilation** by saving groups of trees, which can help funnel breezes while also cooling the air. In some situations pruning up the lower branches to create a "window" between lower branches and the ground will speed up the wind as it squeezes through. (See VI-2 on page 161 and II-4 on page 67 for more about trees.)

Figure VI-1: Whether saved during construction or planted later, this mature sugar maple provides priceless cooling shade and breezes in summer while allowing the welcome low-angle sun of winter to shine into windows.

Figure VI-2: When building a new home, there is hardly any action more important than preserving potential shade-giving trees, and positioning the house in just the right spot to take advantage of this free cooling effect.

❯ **Position patios, decks, or hard-surfaced outdoor gathering spaces** so they are not directly south of the house to prevent the exposed surface from reflecting midday sunlight into windows, and absorbing and holding the heat, which would be released later in the evening and overnight.

❯ **In flood-prone areas, position the building so that its floor level is elevated** well above the surrounding grade. (See VI-5 on page 171 for information about flood-wise landscapes.)

❯ **If you need to create a new clearing for the home, think carefully about how much space you will need.** The process involves being as intentional about the design of your outdoor spaces as you are about your indoor spaces. Plan out the shapes and sizes of yards, gardens, orchards, meadows, patios, paths, driveways and parking, tree groves for shade and habitat, extra space for compost, water storage, etc.—and anything you can imagine wanting. And then make the clearing just large enough for those things. (Engaging the services of a landscape architect or site planner at the start can save you a lot of money and disappointment later on.)

Figure VI-3: Live oaks can provide the classic high shade and airy space below that is so essential to life in the American Southeast.

Credit: SReed

❯ **Avoid creating steep, erodible slopes,** which may be more vulnerable in the future, with more intense and longer downpours. (See VI-5 on page 171 for more about preventing erosion.)

❯ **Plan to add trees to shade hot outdoor spaces.** For more guidance about maximizing the cooling effects of trees, see II-4 on page 67.

❯ **Design path and patio surfaces to be as cool as possible,** by making the surfaces a light color, using paving materials that allow rainwater to soak into the ground, and shading exposed surfaces to minimize the amount of heat they absorb. (More about cool paving in VIII-4 on page 232.)

⊙ **Minimize the amount of imported topsoil needed,** by careful design of the land. Whatever its origin, topsoil that is brought to a new site has a high carbon cost in terms of excavation, transportation, and relocation. Disrupting the smallest area possible will limit the amount of new soil needed for repair. (See V-3 on page 142 for more information on minimizing soil disturbance.)

Last Thoughts

Houses are simultaneously the most permanent, high-impact feature in most domestic landscapes and one of the biggest financial investments most of us will ever make. So positioning a house and designing the whole site to have the lowest ecological footprint is vitally important. The advice presented here helps us protect both the environment and our investment, minimize later regrets, maximize our pleasure and happiness, and anticipate the coming climate realities.

> With careful planning—selecting the right parcel, designing the best access, locating the best house site, and sensitively placing utilities and other exterior uses—you can limit the impact on your property and design a comfortable fit between building and land.
>
> —Mollie Babize and Walter Cudnohufsky, "Building Your Highlands Home," A Highlands Community Trust project of the Trustees of Reservations, 2009

ACTION TOPIC **VI-2** **Fit Landscape to Land** »

Why This Matters

How we situate landscape elements on the ground will strongly affect our carbon footprint. This can be especially true on sloped land, which usually is re-shaped to accommodate our need for level ground and moving around the site (on foot and in vehicles). Large equipment operating and idling, a lot of earth being moved around, materials trucked in to level or shape the ground, topsoil being disrupted or compacted: all of these activities consume large amounts of fuel and release a lot of CO_2 into the atmosphere. If we want our landscapes to have the smallest possible carbon footprint, it just makes sense to design new landscape features so they fit the shape of the land.

Actions

When planning the overall arrangement of the landscape, almost limitless possibilities exist for us to intentionally create gardens and designs that fit the existing terrain as much as possible—instead of significantly altering the terrain to fit the design. This is not to suggest that we shouldn't move earth at all, but rather that the more we can keep earth-grading to a minimum, the more climate-wise our projects will be. General guidelines include:

- Dedicate existing level or nearly level land to the things that require it, such as patios and gathering spaces, car parking areas, playing fields (and, usually, septic drainfields in rural areas).
- Use moderately steep slopes for things that don't need level land, such as multi-story buildings, decks and stairways, paths and terraces, retaining walls, roads, and driveways.
- Avoid or minimize building on the steepest slopes. Anything steeper than a 33% slope (a ratio of 1 foot vertical to 3 feet horizontal) will require some kind of stabilization unless it is bedrock, in which case the land may be difficult, expensive (in carbon terms), or impossible to alter.
- Whenever moving earth near trees, be extra careful not to harm their roots. (See II-1 on page 52 for more about protecting trees.)

Figure VI-4: One effective way to deal with a steep slope is to build multiple low walls and create a gently sloping plant bed (or beds) in between. This can use less stone than building one tall wall, and can be structurally safer.

Credit: SReed

❯ **Limit rock blasting.** To avoid harming a foundation, never blast bedrock near a building or a well. If blasting is required, plan or seek ways to reuse the blasted material on site.

❯ **Design outdoor living spaces to fit into any slope change** in the ground surface. This might mean building a low retaining wall into a hillside to enclose a patio or outdoor sitting space. Or it could mean shaping the ground so it directs water toward a garden needing irrigation or tilts a garden bed slightly upward to be more easily viewed.

Credit: SReed

> **Avoid working on ground that is frequently wet.** If a project area includes (or is near) land that is damp or saturated, or near a pond, stream, or any running water, consult with a wetland specialist, civil engineer, ecological planner, and/or local authorities (Conservation Commission, Natural Resources Commission, etc.) to make sure you are observing all regulations and bylaws; you don't want to accidentally incur a fine or end up having to redo or undo all or part of a project. Some states do not restrict or regulate construction in or near wetlands and water resources, but many do, and it's a good idea to be informed, just in case.

Figure VI-5: When land near a house drops off abruptly, a constructed terrace can enlarge the usable level space.

Even in regions with no wetland regulations, taking care of wetlands makes sense because they provide rich habitat and store large amounts of carbon in their submerged soils. Disturbing wetlands can also result in the release of methane or nitrous oxide, potent greenhouse gases.

Last Thoughts

When we adjust our demands on the land, when we take the time to work with what the site gives us, when we alter our own desires to accommodate the unique qualities of a property, we automatically make a landscape that's more harmonious and fitted to its place than if we impose our every wish on the place. It's really a matter of modifying our preconceived ideas and desires. Even a little bit of this will go a long way toward helping the natural world and letting us make a smaller carbon impact on the planet.

> When you build a thing you cannot merely build that thing in isolation, but must also repair the world around it, and within it, so that the larger world at that one place becomes more coherent, and more whole; and the thing which you make takes its place in the web of nature, as you make it.
>
> —Christopher Alexander,
> *A Pattern Language*, 1977

Why This Matters

We can design every element in the landscape—from paths to steps to walls and patios, from gardens to lawn to orchards and hedges—so that they serve as many purposes as possible. This will enable us to get the greatest value out of every investment of energy and carbon, as well as our valuable time, not to mention the money we spend. Below are several examples of this kind of design thinking; you can probably imagine many more.

Figure VI-6: This dual retaining wall/bench, at the National Museum of the American Indian in Washington DC, perfectly exemplifies multi-purpose design.

Actions

If you intend to build some sort of **low wall around an outdoor gathering area,** either to retain a slope or just to enclose a space, consider making it about 18″ tall, and give it a good solid top, so it can also serve as a sitting wall. If it's a retaining wall, then this edge can also support a raised, easily accessible garden or plant bed above it.

For a front walk, consider designing it so it doesn't just make straight "runway" from one point to another; think about giving it a slightly enlarged area

Credit: SReed

somewhere along the way, perhaps a little nook or mini-landing area, where a bench or a couple of chairs might be positioned. Maybe this spot could offer a nice little view, or the fragrance of a few flowering shrubs. Or maybe it could tuck beneath the spreading branches of a small tree, which itself could also cast some cooling shadow on nearby windows. This little touch will nicely alter the arrival experience for residents and visitors alike without adding much to the cost of the project.

Figure VI-7: A walkway can do so much more than just take people from point A to point B, as this intriguing fountain and courtyard demonstrate.

◉ **If you intend to plant fruiting shrubs,** how about positioning them so they also enclose a space, or provide privacy, or help direct summer breezes toward an outdoor dining area? Alternately, if you wish to plant a hedge or build a fence to conceal an undesirable scene, try designing it as a place for wildlife habitat (with shrubs, vines, perennials, etc.) and perhaps link it to other natural areas, to help provide a migration corridor for species on the move due to changing habitat conditions.

Credit: SReed

◉ **If terrain slopes away from your house** so that there's no place near the house to simply walk comfortably and rest on level ground, think about creating some kind of terrace to provide an area of solidity and "support" for the house; make it big enough for some small (or large) gardens and perhaps also a stone or grassed patio for outdoor seating. Also consider how people will move around within it and/or cross it from side to side, to reach other parts of the property.

◉ **For a garden shed or garage,** could the roof be made large enough, and oriented correctly, to accommodate solar panels, perhaps to power or recharge electric tools and yard equipment? As an alternative, you might add a green roof on the shed or collect rainwater from the shed in a rain barrel for a handy water supply. (See more about green roofs in VIII-2 on page 226.)

⊙ **If you need to support a slope** to keep it from eroding, try including pockets of soil for fibrous-rooted plants to grow, thus assisting with stabilizing the bank. See if it's possible to use local and/or recycled materials, such as boulders from a construction project nearby.

⊙ **When a building's steps lead to a patio or deck,** think about making them generous and wide enough for people to sit on, as an added relaxation spot for social gathering space. If the steps are very wide, you can build a planter box down the center (perhaps with a railing for safety) and use it for growing kitchen herbs.

⊙ **When adding trees near the house** place them carefully so they help cool the building façade or roof during the summer. Small, narrow trees, when planted quite close to the south wall of a house, will shield south-facing windows from the hot overhead sun in summer. Choose native species that will support wildlife with flowers for pollinators.

Figure VI-8: Your lawn could be an orchard!

Credit: SReed

In many parts of North America, a warmer and wetter climate will increase the occurrence of mosquitoes and other unwanted flying bugs. Yet we will still want to be able to dine and have gatherings outdoors, perhaps even more so to enjoy cool evening breezes. One solution lies in **designing a screened porch, gazebo and other structure right from the start,** instead of automatically building a patio or deck, and then later finding that it's uncomfortably buggy and maybe even unusable. Replacing things costs energy and uses more materials, transportation, etc. (For thoughts about controlling mosquitoes, see III-5 on page 92.)

Last Thoughts

This kind of multi-purpose design thinking is the essence of climate-wise landscape planning. Our landscapes are full of opportunities for us to pack extra function and value into every component of the property. In doing so, we will accomplish our goals and have the things we need and want, while simultaneously being as efficient as possible with our use of materials and energy-consuming, CO_2-emitting processes.

This list is only a beginning; we invite you to be creative in your own yards as you figure out your own solutions. The point is not necessarily to do more with less. Instead, the message of this *Action Topic* is to show how you can accomplish a smaller climate footprint by making the most of everything you create, build, transform, and plant in your landscape.

Outdoor spaces are defined by trees, hedges, buildings, and hills but rarely completely enclosed. They are partially bounded, their form completed by the shape of the floor and by small elements that mark off imaginary aerial definitions. Level changes can define spaces and create effects of dynamic movement. A regularly organized space will tilt uneasily if it contains a steep slope. Thus it is safer to make up vertical differences in the approach or in the transitions between important openings.

— Kevin Lynch, *Site Planning*, 1985

ACTION TOPIC (VI-4) Create Energy-Wise Landscapes »

Why This Matters

A basic premise of this book is that our landscapes contain countless big and small opportunities for us to shrink our carbon footprint. Every section of this book addresses this subject in numerous ways—from mowing less lawn to buying plants locally to shading buildings with trees, and many more.

Here, we present concrete ideas for planning the major components of our landscapes specifically to enable us to save energy—whether indirectly in the form of lowered utility bills or directly in reduced use of hydrocarbon-based fuels.

Actions

Action Topic II-4 covers the subject of getting the most cooling benefit from your landscape. In addition to providing shade, though, there are several other ways in which landscapes can help cool a building:

❯ **Cool the air through transpiration.** We can do this by having lots of plants (not only trees) with large leaves near and all around a house. (See Transpiration in "A Primer on Water Chemistry and Plants" on page 75 .)

Figure VI-9: Native bearberry, Canada mayflower, birdfoot violet, and haircap mosses blend with heathers and dwarf juniper (a non-native cultivar) to create an entirely lawn-free yard on this rocky, dry hillside property, which avoids nearly all fossil fuel consumption and CO_2 emissions.

Credit: SReed

▶ **Cool the ground.** Again, lots of plants near a house will help cover bare ground, which can lower soil temperatures by 20 degrees or more. Further, although lawn grass cools the ground better than pavement or pebbles or having nothing growing at all, the soil under lawn grass still tends to be warmer than under plants that have more lush foliage. So it can be a good idea to minimize lawn directly south and southwest of a building. (See Section I for more about making lawns more climate-wise.)

▶ **Also, try not to position hard surfaces like patios and driveways directly south and southwest** of a house. Here is where the sun's heat will either be absorbed and released in the evening and at night, or where lower-angle (afternoon) rays will bounce off the hard surface and reflect into windows, heating the interior. Add shade trees to cool exposed hard surfaces near the south and west walls of a building.

▶ **Build arbors and pergolas** to shade walks and sitting areas and any hard surfaces that absorb the sun's heat.

▶ **Try to create this ideal arrangement of cooling elements:**
- On the east and southeast, position the hard-surface elements like patio and driveway, and plant small to large trees to shade the hardscape.
- Directly south of a house, plant lots of low- and medium-height vegetation. Ideally, have a few small trees near the building, with their canopy high enough to shade windows during midday and allow light and views below the branches, but not big trees whose overhead branches might block sunlight to a rooftop solar array and harm the roof, and/or whose roots might harm the foundation.

The true beauty of the pergola and arbor lies in their countless design variations, from extremes of simplicity to intricacy, and their usefulness in creating a comfortable climate. Their form also provides an armature for growing plants vertically and horizontally in a small space. They increase our awareness of the changing seasons as the vines on them grow, flower, and fruit. Watching the shadows dancing on the ground makes one more aware of the movement of the air from even the gentlest of breezes.

—Chip Sullivan, *Garden and Climate*, 2002

• Southwest (and to a lesser extent southeast) of a building, position smaller trees a bit farther away, to help block the low-angled afternoon sun.

Note: This subject is covered in more detail in II-4 and also in extensively in Sue Reed's book, *Energy-Wise Landscape Design*.

> **Use the landscape to help warm a house in winter,** by not blocking valuable sunlight. To do this, avoid planting tall evergreen vegetation near the house within a cone that's about 30 degrees to the east and west of south, which is the region of greatest solar gain in winter. (To a lesser extent—and this is mainly useful in large open landscapes—on the north side of a building use evergreen groves or hedges to deflect cold winds away from the house.)

Credit: SReed

> **Use energy-efficient maintenance practices and equipment.** It's also important to design spaces, edges, and structures to make maintenance as streamlined as possible, e.g. enable easy passes with mower by creating curves instead of corners; group shrubs and trees in shared beds instead of sprinkled around the property, etc.

Figure VI-10: This 1930s farmhouse is perfectly positioned for energy-savings: a tall hillside on the north diverts winter winds; the house opens to the winter warmth from the south; and a large shade tree to the east keeps the house cool on summer mornings. Small trees near the southwest corner augment the summer cooling effect but don't block low-angle winter sun.

◉ **Water/irrigate efficiently.** Water is an energy issue because the process of getting water from where it is—in streams, lakes, reservoirs, underground seams and aquifers—to where we want to use it, requires effort. And most of that effort requires electricity or power in some form, and sometimes a lot of it. (See III-2 on page 80 for creative suggestions on minimizing irrigation.)

◉ **Whenever possible, use solar power as the source of electricity** for path lighting, decorative lighting, trickle fountains, small water pumps, pool lighting, pond pumps, and other landscape features instead of conventional line voltage. (See more on the subject of renewable energy in VI-8 on page 186.)

Last Thoughts

Although we might not be able to make entire landscapes into carbon sinks (where more carbon in stored than emitted), every small action to reduce our own portion of greenhouse gas emissions is a step in the right direction. Here is an opportunity for all of us to participate in the drawdown of carbon dioxide from the atmosphere. Right in our own landscapes—whether they're big or small, fancy or plain, modern or traditional—we can all contribute to solving the problem of global warming.

> The primary passage of solar heat into a building is through the windows. Walls and roofs also significantly affect interior temperatures. In southern climates, shade is very important to control excessive heat buildup. Since one can't always afford to do everything that would be beneficial, however, one should prioritize starting with the areas that are most effective. The order of importance for shading is: windows, air conditioners, un-insulated roofs, walls next to windows, other walls.
>
> —Ruth Foster, *Landscaping That Saves Energy and Dollars*, 1978

ACTION TOPIC **VI-5** **Design Flood-Wise Landscapes** »

Why This Matters

In the coming decades, many regions are likely to receive more intense downpours, lowland flooding, and flash floods than in the past. Of course, we can't stop major rainstorms, but we can take actions on our own properties that will help reduce harm and destruction, both to our own land and to other lands downstream.

Actions

As mentioned in Section III, one major goal of sustainable landscaping is to encourage rainwater to slow down and sink in, rather than running off or being diverted into storm drains. The following steps toward that goal can be applied in a wide range of landscape types to help absorb stormwater.

⊙ **Let floodplains do their job.** Low-lying land often provides space to hold water that overspills an adjacent or nearby waterway. This is an essential service: it deducts some water from the total volume of flow, and it lets that water spread out and slow down. This reduces the overall impact of flooding on an entire region. Furthermore, it can build soil by depositing useful silt. In such low-lying land, avoid planting species that can't tolerate standing water, and don't build structures that will be harmed or undermined by flowing water.

⊙ In every landscape, **keep the total area of impervious surfaces as small as possible,** and make sure all surfaces are pitched to direct water either toward where it can be used (rain gardens, bioswales, etc.) or to where it will cause the least harm.

⊙ **Create an abundance of vegetated and porous places** to help water percolate downward. This might include rain gardens and/or engineered bioswales, as discussed in Section III. But in some situations a simpler solution could be to shape the ground surface to form one or more depressions or shallow low spots, and to design these places to be both permeable and able to tolerate short periods

Credit: SReed

Figure VI-11: Floodplains are natural depressions where floodwaters can collect and be diverted from the main flow of a swollen river. To protect our own and others' property, we should preserve natural flood basins (even if they're in our own back yards) and let them do their job, instead of trying to eliminate this important storage area.

of inundation or shallow standing water. (*Caveat*: In regions of frequent or extensive flooding, in highly paved areas, and/or where conditions limit the soil's ability to absorb water [clay, bedrock, etc.], this action may not be sufficient. If you have any doubts, consult with a landscape architect, engineer, or other professional.)

Credit: SReed

❯ **Grade the ground surface,** whether pavement or lawn, so that runoff flows as *sheet flow* instead of being concentrated and directed toward outlets. If runoff must be concentrated, consider using a swale—a broad, shallow-sided linear depression that tilts gently toward an outlet point—instead of a ditch or pipe, both of which magnify water's erosive power before depositing it at the endpoint.

Figure VI-12: This perennial bed positioned at the base of a steep slope both benefits from the periodic influx of runoff and absorbs excess waters to protect the yard and house.

Actions to Protect Landscapes from Flood Damage

In low-lying areas, harm from intense rainfall will most likely be from standing water, washout, siltation, saturation of plants, and topsoil erosion. While there is no way to completely flood-proof a landscape, the following suggestions will help minimize damage.

For gardens and plants

- Cover all bare soil, with plants, including lawn, so moving water won't carry away precious topsoil. (*Tip*: Avoid mulching steep slopes as a means to protect bare soil; unless inter-planted with dense cover of shrubs or other deep-rooted vegetation, this mulch will likely be carried away.)
- Garden in raised beds to elevate plant roots out of potentially saturated soil, especially edible crop plants.
- Garden with plants tolerant of brief saturation, in particular plants adapted to floodplains and wetlands.
- Plant trees on a slight mound. It is far better for trees to be planted a bit too high than too low.

- Cover temporarily exposed soil with a cover crop, straw, mulch, or netting of some sort, but nothing that would float away or wash off during high-water events.
- After a flood: if lawn has been under at least an inch of water for more than a week, the weight of the water will have compacted the soil, eliminating the voids that are needed to hold oxygen, and making future drainage/absorption harder. In this case, fork or aerate the lawn. Ditto for garden beds. And if any trees have been standing in water with roots saturated for more than a week, be aware that they might be unstable. Consult with an arborist.

Figure VI-13: In densely built-up neighborhoods, it makes sense to keep stormwater on one's own property, where it can either seep in or nourish gardens, instead of shedding the runoff into storm drains or onto other people's property. This clever grading and walkway demonstrate one homeowner's whimsical solution.

For the built landscape

- As in all construction, shape the ground so that it slopes away from buildings, and outward for at least 15–20 feet if possible (at a minimum slope of 2%) before it rises again, unless other provisions are made to manage drainage.
- Design and build driveways, patios, paths, and other hard surfaces to be slightly elevated (2–4 inches) above the surrounding grade, and laid with a slight sideways pitch so water sheets off. (Make sure the adjoining ground slopes to meet the pavement, to avoid twisted ankles.)

Credit: SReed

> ## Hold Your Water!
> Never direct surface runoff or drain-pipe outlets toward an adjacent property. Even if local laws don't require this, retaining/absorbing runoff within one's own boundaries is still the right thing to do.

Actions to Prevent Slope Erosion

In extreme or extended downpours, some slopes may experience erosion, washout, or even mudslides. The most vulnerable slopes include land steeper than a 33% gradient (a ratio of 1 vertical to 3 horizontal), regions where the soil is very clayey, and slopes with shallow bedrock (especially when that bedrock is also sloping and relatively smooth). However, even short and moderately steep embankments may erode in a deluge.

❯ Where possible, **shape the ground to minimize the steepness of slopes.** This may be done either by lowering the top or raising the bottom of the slope, or both.

❯ To support a steep slope, **build a series of retaining walls or rock embankments.** Or use some form of rock material (ideally local) to armor the surface. *Avoid simply mulching a steep slope* as a means to prevent erosion, because heavy rain will wash away the mulch, exposing the soil below and adding unwanted nutrients to the runoff.

 Note: Whenever building a solid structure such as a retaining wall against a slope, be sure to provide a means for water to escape from behind the wall. This might involve adding loose backfill with a pipe at the bottom that carries water sideways from behind the wall, or leaving "weep holes," i.e., gaps or pipes through the wall, to relieve any water pressure that may build up behind the wall.

 Form terraces that step up the slope, with retaining structures in between. When feasible, and where there is an appropriate outlet area within the site, grade these level areas with a slight cross-tilt, to direct surface water sideways off the slope.

 On a long gentle slope, consider using "contour-swales"—shallow-sided ditches that run across the face of a slope—roughly parallel to the contour instead of angled steeply downhill.

 On steep, erodible slopes that can't be retained or armored, consider using *bio-engineering techniques.* These include vegetation and other non-structural materials set up to fortify the face of the slope. Options include *wattles* (also known as *straw worms, bio-logs, straw noodles,* and *straw tubes*) that are laid perpendicular to the direction of runoff, or seeding the surface with grass, either by hand seeding or hydro-seeding (seed mixed into a slurry that is sprayed onto the surface). For steep slopes, the seed may need to be covered with erosion-control netting.

Donate Your Flood-prone Land?

If your land is located in flood-prone areas, consider donating this land to a conservation organization. Such a donation could ensure that the land remains as part of the larger floodplain in perpetuity, and it might also bring a tax deduction.

Last Thoughts

Flood-wise planning will help preserve landscape plants and minimize soil washouts, and it will protect property and investments. It might also contribute to larger benefits downstream, by reducing the amount of silt and debris carried to other landscapes and overloaded storm drains. A further added benefit will be to assist wildlife by preserving habitat and keeping nearby waterways cleaner.

ACTION TOPIC **VI-6** Install Buffer Zones »

Why This Matters

Technically, a *buffer zone* is an area that keeps two things separated, often created to divide hostile or belligerent forces. Here we use the term in its conservation meaning: *an area of protected land located between a body of water and the adjacent land.* When occupied by natural vegetation suited to the water's edge, buffer zones can be of great value—whether a property abuts a lake, estuary, wetland, river, or even an ocean.

Many properties that contain waterfront, especially bulk-headed (armored) shorelines, have highly managed lawn right down to the water's edge. As a result, lawn clippings and products applied to lawns such as fertilizers and pesticides can pollute the waterway. Further, although the lawn will absorb some amount of rain and runoff, there may also be times when stormwater overflows, carrying pollutants and eroded soil into the body of water, whether it's a lake, a bay, or the ocean.

Buffer zones can help to preserve the shoreline from storm damage and erosion, which will help landscapes better tolerate the effects of climate change. In addition, reducing water pollution and creating shoreline habitat will help support wildlife, especially aquatic species.

Actions

⊙ **Replace turfgrass in the waterfront area** with vegetation that doesn't require mowing. One option is to remove the turf and immediately replant the area, but this approach disturbs the soil, releases carbon, and risks erosion of the exposed soil. A better option is to leave the grass in place and dig holes through it for new plants; but you should avoid mulching in places that could be flooded. The plants you choose for this area should be low care, native woody plants and herbaceous perennials suited for the specific zone and level of salinity. Observe what grows naturally in areas similar to your site. If existing vegetation, (other than invasive plants or turf) is doing well, let it stay. Any buffer is good, but a width of 30 feet or more is recommended.

⊙ If there is a slope toward the water's edge (and if local regulations allow), **create one or more berms parallel to the shoreline** in the buffer area to trap or slow down stormwater flow. The swales

Credit: SReed

Figure VI-14: Instead of having lawn extend right up to the edge of a bulk-headed shoreline, a climate-wise approach is to plant (or allow to grow, or leave in place if one already exists) a strip of natural, absorbent vegetation that doesn't require mowing or synthetic chemicals near the water.

Be Informed about Waterfront Regulations

Many states and cities have laws and bylaws that specify what is and is not permitted within a certain distance from water bodies, potentially including ocean, estuaries, lakes, rivers, ponds, wetlands, and streams. Some areas even regulate how to handle wetland plants, e.g. Florida's mangrove regulations. Penalties for actions that ignore or violate these regulations, even through ignorance, can be severe. Before doing anything that alters your waterfront, find out what's allowed!

Figure VI-15: As with this property on Long Island Sound, many towns mandate that managed gardens be kept a certain distance from a wetland or shoreline. This narrow strip of grass, acting as a transition between the built and natural world, exemplifies what is possible and desirable, even in the most luxurious of neighborhoods.

Credit: SReed

between the berms can act like long rain gardens; and the steeper the slope, the more important they are in protecting the shoreline. (See III-3 on page 84 for more information on building effective rain gardens.) Also, if there is a pathway to the water on a slope, use stepping stones, pavers, or decking to reduce erosion. Many products are available that provide a non-slip walking surface while also preventing erosion on the path.

- If you wish **to preserve the view of the water,** choose mostly low-growing plants with only a few woody or taller plants grouped around the edges to frame the view. But keep in mind that the goal of a buffer is to have enough deep-rooted and wind-tolerant plants to actually help hold the shoreline during a severe storm.

- **Mulch at the waterfront.** Be careful not to use mulch that could float or wash into the water. And never use weed barrier cloth here: this material reduces habitat value by preventing access to the soil where turtles might lay their eggs, toads might bury themselves to stay cool, or solitary bees might make a home.

Credit: GStibolt

Figure VI-16: Instead of extending "landscaping" all the way to the water's edge, we can leave generous expanses of natural landscape, to provide the optimal habitat for aquatic wildlife.

Last Thoughts

Waterfront property owners have a special opportunity to be good stewards of their shorelines. Natural upland and wetland vegetation protects water quality by trapping sediments and taking up nutrients and other pollutants. A buffer area can provide habitat for some wildlife and prevent unwanted or alien species from moving in. Thick buffer vegetation also prevents erosion and stabilizes the shoreline by absorbing wave energy, trapping sediments, slowing stormwater runoff, and moderating the effects of storms and floods.

Of the 121 million acres of coastal wetlands globally, 18 million acres are protected today. If an additional 57 million acres are protected by 2050, the resulting avoided emissions and continued sequestration could total 3.2 gigatons of carbon dioxide. While limited in area, coastal wetlands contain large carbon sinks; protecting them would secure an estimated 15 gigatons of carbon, equivalent to over 53 gigatons of carbon dioxide if released into the atmosphere.

—Paul Hawken, *Drawdown*, 2017

ACTION TOPIC (VI-7) Create Fire-Wise Landscapes

Why This Matters

Fires are natural occurrences in many ecosystems, including prairie, savanna, coniferous forests, and even wetlands. Before European civilization of this continent, relatively frequent fires caused by lightning reduced underbrush and kept those environments vibrant. In fact, ecological research shows that fire itself helps to increase biodiversity within fire-adapted areas. In the last hundred years, we have suppressed fires to protect property. Consequently, forests and their underbrush have tended to become denser. This makes the tree populations simultaneously less able to survive extended droughts and more vulnerable to fires, which have become correspondingly more intense.

Many experts predict that wildfires are likely to increase in frequency and intensity because of higher average temperatures and more severe droughts. In addition, stressed plants may have an increased risk of insect and disease infestations, which may kill some or all of a tree; since standing dead trees don't transpire moisture into the air, total rainfall in a region may be reduced, and this can worsen the fire situation.

The good news is that there are several ways we can protect our property from fire damage. Fire-wise landscaping consists of good planning and care-

Figure VI-17: Fires may become increasingly common in the hotter, drier summers predicted in some parts of North America. Planning ahead could protect property.

Credit: GStibolt

ful management, both of which are described below. Bear in mind that these precautions apply primarily in regions that are especially vulnerable to fire. If you're not sure about your region, ask the local fire department, extension agent, park ranger, or forester.

Actions to Limit the Spread of Fire

Fire-wise landscapes contain three zones of protection. Their dimensions will vary from property to property, so the measurements presented here are general suggestions and should be customized to fit individual landscapes and lifestyles.

Fire Basics

There are three types of wildfires; the differences between them depend on the presence of fire's three basic ingredients: fuel, oxygen, and heat.

- A *surface fire* quickly burns the material above the ground, but below the canopy most trees survive. Fuel includes leaf litter, mulch, shrubs, vines, and other low objects; this type of fire is desirable as a way to prevent the next two types.
- A *canopy*, or *crown fire*, is likely to kill mature trees and will begin only if a surface fire is hot enough.
- A *ground fire*, which burns roots, duff, and other organic materials under the surface, will also begin with a superheated surface fire.

Zone 1

This is the most important firebreak: a 30-foot defensible space around the house and other buildings. Most of the flammable fuel in this area should be removed or isolated. It should also contain a 16-foot wide access and

Credit: SReed

Figure VI-18: This wide band of native buffalo grass provides a protective strip between a Taos, New Mexico, home and the dry sagebrush mesa that surrounds the property.

maneuvering room for fire equipment near buildings. However, you should block access to septic tanks and drainfield areas so heavy trucks don't damage these underground structures. Consider implementing some or all of the following suggestions for creating and maintaining Zone 1.

- Maintain a small area of mowed yard in regions that receive enough rain so that grasses and weeds grow naturally. Choose grasses, clover, or other mow-able ground covers that are well suited to the conditions. Irrigate deeply, but not frequently, so that these plants develop deep roots. In a fire-prone region, dried grass is a fuel for surface fires, so regular mowing and a smart-control irrigation system (which automatically adjusts watering based on input from an internal weather sensor) may be more important here than elsewhere.
- Remove flammable fences, hedges, or other lines of flammable structures that could provide a direct fire link to buildings. Remove dead trees, dead limbs, and brush piles.
- Move woodpiles and compost piles to an area outside of Zone 1. You might wish to use three-sided, dry-stacked cinder block or cement block walls to contain your compost or wood piles with the open side facing away from your house. The cinder blocks could act as a firebreak.
- Under mature trees, remove tall shrubs, vines, Spanish moss, or any other *ladder fuel* that could allow a surface fire to climb up to the canopy. Pay particular attention to poison ivy, poison oak, Brazilian pepper, and poison-wood trees: if these plants burn, the urushiol toxin that causes skin rash is carried in the smoke and can cause severe reactions in people who breathe the smoke.
- Trim the branches of mature trees so that the lowest ones are at least 10 feet from the ground or buildings. For smaller trees, remove the branches on the bottom third of the trunk. This pruning process may take two or three years, because removing more than 20% of a tree at one time could be damaging.
- If shade trees are desired, use deciduous trees instead of the more flammable conifers.
- If there are mature pine trees or other resinous plants in this zone, trim the branches to 20 feet from the ground, and if possible, remove enough

conifers so the crowns are 10 to 15 feet apart. Remove most small conifers, and don't plant new ones.

- Planting areas should be widely separated from each other to provide breaks in the fuel for a fire.
- Check local forestry or extension departments for a list of local flammable plants to avoid.
- Remove leaf litter, pine needles, and other flammable items from the ground and from roofs and gutters regularly, especially when danger of fire is high.

Special considerations within six feet of the house include:

- If this area isn't covered with lawn, use bricks or other nonflammable stone or cement pavers to cover the soil. Any type of vegetative mulch could burn even if it is damp, so avoid using it here.
- Pavers are a fire-wise choice for outdoor living areas; avoid wooden decks. The open space under a flammable deck provides a pathway for fire. If wood decks need to remain, treat them with fire retardant and use metal

Credit: GStibolt

Figure VI-19: In Fire Zone 2 (i.e., more than 30′ from buildings), removing low branches and shrubs can keep a ground fire from sweeping through a landscape.

grating to skirt the undersides to prevent flammable debris from blowing under the deck.

- Remove tall shrubs, vines, trellises, or other flammable objects from under eaves. They could serve as a fuel ladder to elevate a surface fire directly into the attic space.

Zone 2

Within 30 to 60 feet of buildings, moderate thinning and fuel reduction should break the continuity of fuel available for a wildfire. Remove all poison ivy and related plants, and most dead plant material. In this zone it may be acceptable to leave some of the standing deadwood for its wildlife values, but remove any ladder fuel around them. Remove ladder fuels from under mature trees—particularly conifers. Clear out some vegetation between trees and plantings. If the access road or driveway runs through this zone, treat it as Zone 1 and thin back plants to 15 feet on either side.

Zone 3

Within 60 to 100 feet of buildings is the transition to the wild areas. Maintenance should include removal of poison ivy, and ladder fuels under conifers and dead wood.

Actions to Defend Against Fire

Because fires in arid locales may sometimes get out of control, it's important to provide some line of defense. Having ready access to stored rainwater can be an important tool in this effort.

▶ **Collect rainwater from your roofs and other impermeable surfaces into a cistern for firefighting.** A 2,500-gallon tank is a good size, but a larger tank might be necessary if this water to be used for irrigation as well. Provide access that is large enough for a large fire hose (but too small for a child to fall though) and create an obvious label so firefighters can spot it. If the top is open, include a screened cover to prevent mosquitoes. Plan for a spillover from the cistern into a low area on your lot or into a pond. Check with your local authorities to find out if there are regulations for cisterns.

⬤ **A pond, water garden, rain garden, or bog garden** may be a good addition to a fire-wise landscape because the water-rich plants supported by these wet environments are not as flammable as others and the water itself could provide a firebreak. A large pond could substitute for a cistern and could be used for fire fighting around your property.

⬤ **Remember to plan for fire equipment access.** The cistern and water features should be either close to buildings or toward the far edge of Zone 1 and probably away from the septic system as well. (See Section III for more information on rain gardens and ponds.)

Actions After a Fire

⬤ **If the fire was a quick surface fire,** the roots of grasses and shrubs may be viable and will start to sprout within a week or two. If this is the case, rake away the debris and excess ashes (it's okay to leave a thin layer to break down), water remaining plants, and remove charred, dead wood from Zone 1 so it doesn't become fuel for the next fire. Any ashes you have removed will be a good addition to your compost pile. Mature trees will usually survive a quick fire.

⬤ If the fire was hotter and longer lasting, the soil may have been overheated. Dig several inches into the soil in an area near where perennials or shrubs were growing. If the small roots are white and turgid, they are probably viable. In this case, leave things alone and wait to see what grows back. If the small roots are blackened or mushy or if the soil is dry and hot, it has probably been cooked. You'll need to develop a plan for seeding the yard area and replanting the gardens. Consider planting some inexpensive annuals to fill in the gaps while waiting to see what will grow back.

Last Thoughts

Fire-wise planning can help landscapes survive the more frequent fires that are predicted as one possible result of climate change, particularly in arid regions. Reducing local fire damage may also help mitigate climate change and help wildlife survive by preserving habitat. But remember that some wildlife depends on fire to develop the habitat it requires, so fire suppression is not always the best strategy everywhere. If you own large tracts of land in fire-prone regions, check with a land manager or forester about how best to balance the occurrence of natural, necessary fire against the need for appropriate fire suppression.

Additional Resources

- Local forestry organizations should have information tailored for each region.
- National Fire Protection Association, a global nonprofit organization provides materials and training for their Fire-wise Communities program: firewise.org
- A description specific to the California problems with dying trees and increasing wildfires is provided in "An Astounding 102 Million Trees Are Now Dead in California," by Brad Plumer, Nov. 22, 2016, accessed Aug. 8, 2017 on vox.com.

ACTION TOPIC (VI-8) **Incorporate Renewable Energy** »

Why This Matters

While this book presents many ideas for how to reduce our carbon footprint by consuming less energy in our landscapes, this *Action Topic* shows we can generate some or all of the energy we need, to potentially replace some of the energy that might otherwise come from fossil fuels.

In addition to the coming transformation of large-scale transportation and energy infrastructure, many new energy technologies are also emerging that work well at small scales. The more we can generate our own electricity from clean renewable sources, the more we can reduce our own small share of greenhouse gas (CO_2 and methane) emissions resulting from conventional coal- and oil- and natural gas-based electricity production.

Renewable energy is a landscaping issue because all three of the options presented here are affected by major elements in the landscape such as trees, gardens, and buildings.

Note: Before investing in or erecting any of the structures presented here, check local regulations to make sure the system you are considering is allowed and/or properly positioned in your location. Step two in every case is to check with local providers to find out about feasibility, costs, and whether any rebate programs exist.

Solar Electricity

For most individual property owners, solar electricity is the most feasible and affordable source of renewable energy. Solar panels last a long time and are continually becoming more efficient. When homeowners think of installing a solar energy system for their home, rooftop solar is likely to be what they have in mind, and for good reason. Installers have developed many standardized methods that make most systems fairly easy to build. In the years ahead, roof shingles made of solar cells may become not just available and affordable, but the norm.

However, there are also many good reasons to install solar on the ground instead of on the roof. In some cases, a roof is oriented the wrong way relative to the sun—in general, a roof needs to face within 15–20° of true south—or a roof might be too steep, weak, or otherwise structurally incapable of accepting a system. If this is the case, or if a homeowner is reluctant to build something on the roof or alter the building in this major way, a ground mount system can be the same price per watt to install, and even more efficient than

Credit: SReed

Figure VI-20: Pole-mounted solar arrays are an increasingly popular choice, especially when a house's roof orientation will not provide sufficient solar gain.

a roof-mounted system. In order to function effectively, some rooftop systems may require removing large trees that would otherwise provide valuable shading and cooling for the building, along with all the other benefits large trees give us.

There are two general ways to go: a frame-mount, in which a metal grid on the ground holds panels at a fixed angle (which might be adjustable for seasonal changes), or a pole-mounted system, which can have a tracking system installed to rotate the panels so they capture the optimal sunlight at all times of the year.

Wind Power

Although windmills in one form or another have powered civilization since at least 500 AD, we are currently experiencing an unprecedented expansion in new forms of wind energy generators. Massive wind towers, both offshore and on land, can now generate up to 6 million kilowatt-hours in a year, enough to power 3,000 homes, each.

For the purpose of individuals generating electricity, wind turbines come in an array of lesser sizes and creative designs, including mini- and micro-turbines suited for household use. And many new designs focus on increasing efficiency by replacing the traditional spinning blades with clever fins and other compact or vertical-axis systems.

Wind energy can be a great solution in the right situation. Wind turbine manufacture and installation do require some fossil fuels (as do solar panels), but this is a minimal cost in the long run of operating free of all fuel costs. The undesirable effects of large wind farms are dramatically reduced when we consider smaller systems and innovative

Credit: Sally Naser

Figure VI-21: In this vertical-axis turbine, stacked airfoils catch the wind like airplane wings, and the resulting lift causes the central rotor shaft to spin. This turns a generator that converts the wind's energy into electric power.

designs, such as vertical-axis systems, which eliminate vibration and noise. And wind turbines work well in the dark and in storms, when solar can't do anything.

To be effective, all wind turbines require sufficient direct, uninterrupted wind. And that's where the landscape comes in: the turbine has to be set up in a location where prevailing winds are not blocked by trees or buildings. So its potential location must be an integral part of all landscape planning.

Actions

Consult with the local utility to find out whether you can (or must) feed electricity onto the grid, and about any regulations affecting that option. Check online to find potential wind speeds in your region. Get a wind evaluation to make sure buildings, trees, or other tall structures won't block or break up the wind currents you aim to use. This is also a landscape planning issue, as large trees that provide shade and other important benefits might be a drawback relative to wind. Check with local authorities to make sure zoning regulations allow wind turbines, and if so whether a tower's height is also regulated. Most zoning requires some minimum distance between any tower and your boundary lines.

Geothermal Heating/Cooling

Geothermal heating and cooling systems transfer the generally stable temperature of the ground into a building, which helps reduce the amount of electricity and/or fuel needed to supply heat or air conditioning. The ground itself is an insulator that holds the sun's heat. As a result, the temperature below the surface, at depths that vary by region, remains relatively steady at 50–60°F.

There are two types of geothermal systems:

1. *Open-loop systems*, which draw water from a vertical well and pass this water through the heat pump before circulating it back to the ground. For this to work, the site needs to be accessible to a well-drilling rig, and groundwater needs to be within a reasonable depth of the surface.
2. *Closed-loop systems*, which rely on tubing laid in a continuous slinky-type of coil installed in long horizontal trenches. In this case, the fluid that carries the ground's heat into the heat pump is antifreeze instead of water; it moves in a self-contained and constantly recycling loop.

If a site allows such a horizontal system, it is easier and less costly to install than the drilled well and open loops. On the other hand, a vertical well occupies less space, can be more efficient to run, and does not interfere with the placement of large trees in the landscape.

Credit: SReed

Actions

Contact a heating/cooling professional to evaluate whether and which type of geothermal would work in your situation. Hold off on designing major landscape features, structures, gardens, etc. until the geothermal system has been located. Both trenches and wells might affect how a landscape gets arranged.

Last Thoughts

One of the best ways to shrink our carbon footprint is to just stop using conventional carbon-based fuels. If you plan to use your own renewable energy sources to power your household, you'll need to plan your landscape design around them. If you're not sure what you will do, consider installing temporary fixtures such as arbors or trellises along with herbaceous plants in the area until the plans are set. And you can create some butterfly or pollinator gardens while you wait!

Figure VI-22: Geothermal piping, when laid horizontally, can occupy a substantial portion of one's yard. It's important to plan all landscape construction and plantings around this permanent but easily forgotten feature.

ACTION TOPIC VI-9 Design Climate-Wise Driveways and Parking

Why This Matters

Almost every square foot of surface that we drive and park on has to be formed for that specific purpose. Otherwise, the ground alone would prove too soft to handle the weight. Unfortunately, this process of creating hard surfaces almost always involves tons (literally!) of materials being mined or blasted or somehow extracted from the ground, loaded into large trucks, transported to the work site, dumped, poured, or shoved around by heavy equipment, graded, shaped and compacted...all to make the ground strong enough to carry vehicles. In addition, paving materials such as asphalt and concrete require vast amounts of energy—mining, trucking, heating, mixing, trucking again—to produce and deliver. The whole process of creating pavement is energy-intensive in the extreme.

Huge amounts of land are already paved. The lower 48 states contain about 4 million miles of road, and this figure doesn't include parking lots, driveways, and other paved areas. Data for the exact amount of energy expended—and CO_2 emitted—in the creation of pavement are extremely elusive, but one thing is clear: designing our car zones thoughtfully will help limit their overall climate footprint. The following guidelines may be used both in new construction or when renovating/redesigning old paved surfaces, to minimize our impact.

Credit: SReed

Figure VI-23: Pavement can be one of the hottest features in a landscape, so every action to help shade the surface, such as the small spreading trees of this Florida landscape, will make a big difference.

Actions

▶ **Make paved areas big enough**—but not too big, and in the right shape. Here's how:

- Twelve- to 14-foot-wide driveways tend to be the norm, and driveway contractors might try to insist on this, but a driveway that wide isn't always necessary. Cars are about six feet wide, trucks a bit wider. A modest driveway can be nine or ten feet wide, more if it's long and cars need to pass or if the driveway will frequently be used by large vehicles. Tiny urban driveways are often just eight feet wide. The point is to assess every situation and aim to make the pavement no wider/bigger than necessary.

- Make turning radii 15 or 20 feet in diameter. Except for tiny cars, most vehicles can't turn more tightly than this. Larger vehicles might need a 25–30-foot turning radius. Extra turning and backing consume more gas and wear the surface, but too much pavement is both wasteful and unattractive. (However, if your driveway is very long, think about how fire trucks would turn around if they were ever called to your house. And if you run a home-based business that requires delivery/pickup services, these vehicles will need a larger turnaround space.)

- Make parking spaces big enough to use easily (minimum 8′×18′) but no bigger than necessary (10′×20′). The exception is spaces for wheelchair

Figure VI-24: Apart from shading a driveway, light colors and granular paving materials are the most effective way to keep the surface cool. Furthermore, using locally available materials helps shrink our carbon footprint.

Credit: SReed

vans, which need to meet standards set by the Americans with Disabilities Act: in most cases, these parking spaces must be at least 11 feet wide, and adjacent to a five-foot wide access aisle (which may be shared with a second parking space).

❯❯ **Design the paved area so it works right** and serves its purpose—so it won't need to be changed later. The goal here is to create paved surfaces that simply and unobtrusively do their job, without creating uncomfortable or dangerous situations.

- Position driveways and parking in the right place. Think carefully about the arrival experience created by a driveway; if possible, give people in arriving cars a glimpse of the house entrance, but then direct the cars away from the building, to park somewhere nearby but not dominating an important view from inside the house.
- If possible, let driveways be shaded by either buildings or trees. This extends the lifetime of the pavement and also keeps cars cooler, saving on vehicle air conditioning and making life more comfortable.
- Make the paved areas a comfortable slope. Although highways and primary roads rarely exceed 7–8% slope, driving surfaces can be as steep as 10–12% without causing problems for everyday use. Any steeper than this and the

Credit: SReed

Figure VI-25: These adjacent Southampton, New York, driveways demonstrate two beneficial but contrasting approaches to climate-wise paving. On the left, local and granular materials keep the driveway's carbon footprint to a minimum, but the light colored granite blocks on the right, probably neither local nor easily manufactured, keep the driveway cooler.

surface can be slippery in winter and/or eroded if unpaved. (While it's true that driveways in mountainous areas might be as steep at 20%, this slope is not generally recommended.)

- Make parking areas and other places where vehicles need to park and turn around no steeper than 5% pitch in any direction, but not flatter than 1–2% overall (unless subsurface drainage inlets are provided), so that water can drain off and not cause puddles or frozen patches. (Any more detail than this gets us into road engineering, which is beyond the purview of this book.)

⊗ **Choose the right surface material.** The best choice depends on many factors, including microclimate, site conditions, the size of the paved area, and local custom. See Section X for a more detailed discussion of materials used in construction projects.

In certain situations, particularly parking lots, it can make sense to make pavement porous or permeable to help reduce stormwater runoff and also to help keep pavement cool. If this idea makes sense in your situation, consult with an engineer or landscape architect who is experienced in this work. (See Item VIII-4 on page 232 for more about cool pavement.)

Slope Calculations

Slope in landscapes is generally represented as a percentage; i.e., how many feet of vertical rise occur within 100 feet of horizontal distance. So a 5% slope represents 1 foot of vertical rise in a 20-foot horizontal run (or 5 feet vertical in 100 feet horizontal). A 10% slope is 1 foot of vertical in 10 feet of horizontal; a 25% slope is 1 foot of vertical in 4 feet of horizontal, and so on. Various methods for determining the slope of your own driveway or land may be found by searching the internet.

Pavement and Trees

If a driveway or parking lot must be built near a tree and/or within the root zone of any trees, be sure to prepare for construction by cleanly cutting (not tearing with a backhoe!) the tree roots. Then generously water the tree (about one inch per week) to help the roots heal and grow. Ideally, do this work several weeks or months in advance. Any work within 5–10 feet of a tree trunk should be avoided; if it is unavoidable, you should be guided by a certified arborist. (See II-1 on page 52 for more about taking good care of trees.)

Last Thoughts

Many personal vehicles are becoming become smaller, lighter, and electric-powered. Yet, plenty of people are still driving large pickup trucks and over-sized vans. Construction standards might (or might not) evolve somewhat to recognize the less-demanding requirements of smaller cars, and this is something to be hoped for. But at the very least, every parking area, whether for a single home vehicle or for a lot full of cars, will likely soon need to include electric charging stations.

No matter what kind of driveway you have, whether it's surfaced with sand, stone, asphalt, concrete or a nice new batch of crushed shells, every cubic yard of excavation, every truckload of fill, every ton of pavement, and every square inch of area equals energy spent on that driveway. If efficiency is your aim, an essential first step is to keep the car zone to its minimum workable size.

Sue Reed, Energy Wise Landscape Design, page 184

HERBACEOUS PLANTS

Overview

While woody trees and shrubs might be considered the backbone of the landscape, often it is the herbaceous plants (i.e., those with non-woody stems) that set the tone or theme of a place. Herbaceous foliage provides texture and habitat, while the flowers add color and beauty, and provide the nectar and pollen that many organisms depend on for food. Herbaceous plants also play an important role in the biodiversity and resilience of an ecosystem; their roots contribute to the complexity of soil ecosystems, and their foliage is a major factor in shading the ground, which helps preserve vital soil moisture.

There are many ways to classify herbaceous plants: by life cycle, growth habit, and niche/use in the landscape. These categories represent perspectives of different groups of people who deal with plants, including botanists, horticulturists, ecologists, conservationists, designers, and planners. All terms are important for gardeners to understand.

Life Cycles:

- *Annuals* complete their lifecycles in a year.
- *Biennials* usually require two growing seasons in order to complete their lifecycles.
- *Perennials* live for three or more years—many live for many years. Most perennials die back at some point during the year, but some are evergreen.

Growth Habits:

- Forbs are herbaceous flowering plants that are not graminoids. Most bulbs would be included in this group.
- Graminoids (grasses, sedges, and rushes) are flowering plants. Some are bunching, while others spread via sprouts and root-like stems.
- Ferns and fern allies (whisk ferns, horsetails, spike mosses, clubmosses, and quillworts) are primitive, non-flowering plants.
- Vines are climbing or twining: some reach up with tendrils and some twine anything upright. Some herbaceous vines die back each year, and others persist without growing woody. (However, many vines are, indeed, woody).

Landscape Use:

- *Ground covers* include both spreading and bunching plants that grow to no more than 1 foot or so tall. This category doesn't include lawns because they require regular mowing to stay low to the ground, but most lawn substitutes are ground covers.
- *Meadow plants* vary widely, but the unifying characteristic is that they can sustain themselves with very little care. Most meadow plants will grow to less than 4 or 5 feet tall.
- *Butterfly garden plants* support butterflies, moths, and other pollinators by providing nectar or pollen for adults and food for the larvae—the caterpillars.
- *Border plants* are those that can mingle with other plants. Most borders are built with perennials so plants can grow together for a long-lasting, semi-natural landscape feature.
- *Climbing plants* can be used in several ways, from covering trellises or pergolas to helping to create a denser habitat in the wilder sections of the landscape.

In This Section

The *Action Topics* in this section include ideas for successfully using herbaceous plants in the landscape, ideally with minimal irrigation and disruption of the soil, minimal use of fossil fuels, and maximum support for nature and wildlife.

- VII-1: Choose Regionally Appropriate Plants

ACTION TOPIC **VII-1** Choose Regionally-Appropriate Plants

Why This Matters

"Choose the right plant for the right place." This simple and oft-repeated advice has long been appropriate for gardeners and landscapers in every region, but it's even more important now, as climate uncertainty is added to the mix. Here's why:

- Extended droughts and water restrictions are more likely in the future, but we can plan ahead with smarter plant choices and water-conserving planting arrangements.
- Severe storms may also be more likely in the future, so planning for too much water can be as important as planning for too little.
- Pollinators and other wildlife may struggle with changing weather patterns; some plants will respond (and already are) with earlier-than-normal blooming cycles. Our pollinator-friendly landscapes can provide the support they need.
- When we choose appropriate plants and use good planting techniques, the plants live longer. When plants live a long time, the soil is less disturbed and can continue to store its carbon.

Actions

❯ **Save time and money by planting perennials instead of seasonal annuals.** As discussed in V-2, the common practice of planting thick borders of annuals (such as impatiens, pansies, petunias, and begonias) in full bloom several times a year has some unintended negative consequences:

- Repeated planting and pulling up of seasonal plants disturbs the soil, which releases CO_2 and disrupts the soil's ecosystem.

Figure VII-1: This typical display of plants for sale includes mostly annuals in full bloom, which means that the plants have completed their life cycles and, soon after being planted, will likely fade or die. Later, when new annuals are added to "refresh" the landscape, this disturbs the soil again. A more sustainable practice is to use long-lived perennials wherever possible.

Credit: GStibolt

- Many common annuals are not native to the region where they are planted, so they may not provide the same ecosystem services that a native plant would, such as edible foliage, pollen, and fruit at the right time of year.
- To keep these plants looking good on the shelf, many growers and distributors use systemic insecticides, mostly *neonicotinoids*, which can be extremely harmful to bees and other pollinators. See "A Primer on Landscape Chemicals" on page 19 for more information on neonicotinoids and other pesticides.
- Some of the plants commonly used for seasonal plantings require large amounts of water over their season—impatiens, which are so popular, are among the thirstiest.
- The production and transportation of these plants to garden centers has its own carbon footprint.

Many commercial and municipal landscapes are maintained with annuals because this convention is written into contracts. But as we move into the future, we can rewrite these contracts and replace short-term plantings with dense perennial borders that need less care, save money, and are more climate friendly in the long run.

Note: If you crave winter pansies or summer petunias, consider growing them in containers; this will leave the soil in the landscape undisturbed.

⊙ **Choose native, drought-tolerant perennials.** (See IV-1 on page 105 for a definition and discussion of native plants.) Plants native to a region are already adapted to normal variations in local weather and may be more likely to tolerate some new patterns of rainfall and temperature. Because perennials last for more than three years, they have a better chance of becoming established with deep or wide-spreading roots, to better withstand both droughts and erosion. Also, their longevity means that the soil around them will be less disturbed and better able to retain its stored carbon, its structure, and its ecosystem.

For general landscape plants, choose those that are known to be drought tolerant, so that once they are established, they'll be able to better survive extended dry periods. For areas around water features where water is always present, choose plants that are adapted to moisture. However, keep in mind that rain gardens can dry out during droughts, so rain garden plants should be able to withstand flooding *and* droughts, particularly in areas that have a "dry season" each year. (See Section III for information on both drought-tolerant landscapes and rain gardens, and other water issues.)

Visit local parks and conservation areas, and if possible, talk with the rangers or caretakers about how native plants arrange themselves and work together. Local native plant societies may also offer advice, guided field trips, and perhaps even native plant sales. For information, see plantsocieties.org.

Note: You could also add native, drought-tolerant, and self-seeding annuals to the mix if the randomness of their locations in following years is appropriate for the location, and if not too many would need to be pulled out or transplanted to new locations. Self-seeding annuals require the soil to be disturbed when planted, but they add diversity to the ecosystem.

⊙ **Near-natives or non-natives may be climate-wise plant choices.** In some cases, a *nativar*, which is a native plant that has been selected for certain characteristics such as dwarfing or slow growth, may be a reasonable choice

Figure VII-2: Try to attend and buy from native plant sales, which are usually sponsored by local native plant organizations, whose members can provide expert advice on choosing the best plants for specific locations.

Credit: GStibolt

Credit: GStibolt

Figure VII-3: Zinnias can make a good addition to butterfly gardens. These flowers are non-native annuals, but they are grown from seed rather than as full-grown plants, so including them doesn't require digging or disturbing soil.

for a small landscape space. *Caveat*: Quite often, nativars are clones of one individual plant that developed the chosen characteristic (perhaps through a genetic mutation). Repeated replication of this one variant (for commercial production) means that this plant will not offer the genetic diversity of a true native. It also might not provide the same habitat/food value to pollinators and other wildlife, and it might alter the gene pool of the local natives.

However, non-native plants are widely sold and some may offer an easy-to-grow, drought-tolerant choice. Clovers, zinnias, and marigolds are examples of non-native plants that could add variety to a pollinator garden. Choose plants that do not tend to naturalize (reseed or otherwise spread on their own) and are not invasive. This is a complex issue that merits careful thought and weighing of pros and cons. See VII-3 on page 210 for more on invasive plants.

▶ Use planting methods to achieve the highest survival rate. For most situations, it's best to plant perennials in groupings or drifts because they will create a better environment when growing together. It's not as important to allow for the adult size of perennials as it is for trees and shrubs, because when perennials become overcrowded, they can be divided and moved to new spots in the landscape. In addition, a close placement will help minimize weeds and the need for mulching.

▶ Choose young plants. Plants that are still in flats or in 4-inch pots will adjust to the landscape more quickly than larger plants. Group the plants in drifts with slight swales between them so water can seep into their root zones.

▶ For larger container-grown perennials, prepare the planting holes so that the plants will sit at the same level in the ground as they were in their pots. Spread

Plants Adapted to the Hottest Conditions

Most plants use a method of photosynthesis called *C3*; in this process CO_2 is absorbed into the leaves through their *stomata* (pores). During hot and/or dry conditions, C3 plants close their stomata to conserve water, thereby halting photosynthesis.

In contrast, *C4* plants have developed a different chemical pathway so that photosynthesis can take place even while stomata are closed, giving them a great advantage on hot days. C4 plants (including crabgrass, corn, and sugarcane) are more drought tolerant and can grow faster when others have to slow down due to the heat.

And finally, CAM plants employ a different strategy: they only open their stomata at night when it's cool, when CO_2 is absorbed, and they use both C3 and C4 pathways for photosynthesis at different times of the day and night. CAM plants are mostly succulents and cacti, including members of the pineapple family (bromeliads and Spanish moss), the agave family, and even some orchids and geraniums.

In hot climates, choosing C4 and CAM plants can improve gardening success. "Both the C4 and CAM pathways have evolved independently over two dozen times, which suggests they may give plant species in hot climates a significant evolutionary advantage." (Source: "C3, C4, and CAM plants," www.khanacademy.org)

the roots out and work soil back into the planting holes so the roots splay out as far as possible. Flood the planting holes and press the soil into place so that there are no large air spaces. Build a watering saucer with a ring of soil around the plants so all the water from irrigation or rainfall soaks into the soil around the root ball and does not run off.

Unless there is a heavy rainfall (i.e., more than half an inch of rain), additional watering is needed to help the new plants become established, even if they are normally drought-tolerant (except for cacti and bulbs, which should only be watered sparingly after the original planting so they don't rot).

⊗ **Plant seeds in a controlled environment** such as flats or starter pots to give them the best chance for survival, or sow them directly in the garden or meadow to provide a more naturalized arrangement. If sown in the landscape,

Figure VII-4: The High Line Park in New York City was built on an abandoned elevated rail bed and planted with native or near-native plants. Now providing a vibrant wildlife corridor as well as a place for people, the park is opening people's eyes to a new definition of beauty in landscaping.

Credit: GStibolt

they need a light mulch of straw or pine needles to help hold in the moisture and to hold the soil in place (seeding on slopes requires erosion control).

Whatever strategy you choose, it's important to learn what the seedlings look like so you can distinguish them from other plants. Also, as you tend your landscape, you can be on the lookout for desirable seedlings growing wild.

Last Thoughts

As more and more public gardens—such as New York City's High Line Park and Chicago's Lurie Garden—are being planted with native species, they are changing our perception of what a beautiful landscape can look like. These gardens are welcoming to pollinators, birds, and other wildlife. In addition, more homeowners, communities, schools, and municipalities are adding natives to their landscapes, creating whole neighborhoods that will be more resilient in the face of climate change.

> Public gardens show people that native plants can be beautiful. They elevate the plants by growing them in a variety of design contexts, from naturalized settings to formal gardens. It changes people's perception.
>
> —Sarah Beck, program manager for the American Public Gardens Association

ACTION TOPIC VII-2 Garden to Support Pollinators »

Why This Matters

Pollinators are the tiny minions of the natural world. They fertilize many food crops for wildlife and humans, so they directly support our own survival. They provide a mechanism for plant diversity, via cross-pollination, which may help plants survive climate change (see "How Does Adaptation Work?" on page 61). And they play important roles in local ecosystems—as predators for other insects and, perhaps most important, as prey (i.e., food) for other organisms. Insects are particularly essential for birds—both local and migratory—who need insect protein to rear their young. In fact, insects form the base of most of the food webs that keep nature functioning, which is an especially important job as the planet warms and many species (including ours) are forced to make adjustments.

Pollinators are quite vulnerable to our human interference in nature, because they:

- Are at risk from the use insecticides used both in commercial agriculture and in producing plants for retail sale (see "A Primer on Landscape Chemicals" in Section I).
- Struggle to find larval food sources due to the use of herbicides in commercial agriculture.
- Are frequently killed by insecticides applied as part of conventional lawn treatments. This especially affects many native bees that build their nests in the ground. *Note*: These solitary (non-hiving) bees are not aggressive and will only sting if handled.

Actions

❯ **Choose a good spot for a pollinator garden.** Here's how:
- Find a mostly sunny spot with at least six hours of sun a day.
- Locate pollinator gardens near pre-existing windbreaks.
- Provide nearby shelter for pollinators to avoid predation.
- Build pollinator gardens near crop plants so pollinators stay in the area even when the crops are not blooming.

Co-evolution: Here's How It Works

Plants and their pollinators have evolved together over the millennia. Plants supply nectar and pollen, which attract pollinators—butterflies, moths, bees, wasps, ants, or hummingbirds—which move from plant to plant. Pollen grains from the male parts of one flower are transported to another flower where, if all goes well, they land on that flower's female parts. And voilà: the next generation begins! This cross-pollination ensures genetic diversity in the species (for more on biodiversity, see IV-4 on page 117). Approximately 80% of flowering plants depend upon animal pollinators. In turn, many pollinators depend upon the plants for their food.

Another aspect of co-evolution is the fact that many plants tolerate being chomped on by various herbivores including caterpillars, the larvae of butterflies and moths. Some plants have developed toxins to discourage predation, but various insects have evolved with digestive systems that tolerate those poisons. A good example of this type of co-evolution is milkweed and monarch butterflies; the larvae of this butterfly feed on (and depend on) milkweed foliage that is toxic to many other insect species, but more important, both the caterpillars and the adults are then toxic (or at least taste bad) to birds and other predators.

Figure VII-5: This (European) honeybee is just one of countless bee and insect species, most of them native, that will visit and pollinate beebalm flowers. Beebalm is an excellent addition to pollinator gardens in all of the eastern US and into Canada.

Credit: SReed

Figure VII-6: Monarch butterflies can feed on nectar from milkweed (as shown) or many other flowers. Their larvae, however, can feed only on milkweed plants. Many people garden to support these easily identified butterflies, but be sure to include larval food sources for other butterflies as well.

Credit: GStibolt

- Locate pollinator gardens near wilder areas with stick piles and unplanted and un-mulched areas to provide shelter and nesting areas.

Plant a wide variety of mostly native flowers:

- Include as many types of flowers as possible to attract all types of pollinators, from tubular flowers for hummingbirds and long-tongued insects to wide flower heads, such as those in the daisy family, so butterflies have good landing spots.
- Plant large groupings of each plant species to create large "pollinator targets" that pollinators—even those with relatively poor eyesight—can see.
- Choose plants that flower at different times of the year to accommodate pollinators that may emerge early or late due to temperature anomalies.
- Use plants that provide both nectar and pollen sources.

Plant larval food plants. While flowers and adult insects often get the most attention when people design pollinator gardens, food for the larvae (caterpillars) may be even more important for attracting butterflies and moths. After all, it's the larvae that become the next generation! For example, the butterfly bush (*Buddleia davidii*) may be quite effective at enticing butterflies to a garden, but *its foliage is inedible to most butterfly larvae*. In addition, it's native to China and has been shown to be invasive in many areas of North America.

Some butterfly larvae can survive on only one type of plant. The butterfly larval food that has been receiving the most attention recently is milkweeds for monarchs. Milkweed used to be much more common, but much of it has been eliminated near commercial agriculture fields that are treated with herbicide. Since most monarch butterflies in North America migrate to a small area in Mexico, the reduction in their numbers has generated more interest than for others species. (Plus, they are easily recognized and admired for their

Figure VII-7: Even though butterfly bush has been proven to be invasive in many areas, it's still being widely sold in nurseries. (Some gardening traditions die hard!) Even worse, it does not support the larvae of adult butterflies that enjoy its nectar. Climate-wise butterfly gardeners choose plants not only for their flowers and nectar, but for their ability to feed butterfly larvae, so the whole life cycle is supported.

Credit: GStibolt

beauty, unlike thousands of other equally valuable, but plainer, butterfly and moth species.) However, there are many other butterflies and moths, and they all need larval food sources, too. Learn what butterflies and moths need and help them by supplying their larval food sources.

Dedicated butterfly gardeners cheer when caterpillars eat their plants, because this means that their landscape has become part of the local ecosystem!

❯ **Avoid poisons—even if organic.** Landscape-wide applications of pesticides, especially those that are regularly applied by lawn service companies, are harmful and wasteful for many reasons, as discussed in I-1. Attracting butterflies and other pollinators to the landscape is one of the best reasons for a poison-free environment.

Neonicotinoids are systemic insecticides that become infused throughout the entire plant, including the nectar and pollen. (This is true even if just the seed was treated.) See "A Primer on Landscape Chemicals" in Section I for more details.

Native Bees and Wasps Are Pollinators, Too

Butterflies and European honeybees have received most of the attention when it comes to supporting pollinators, but our native bees and wasps play important roles in local ecosystems. In North America, there are about 4,000 species of native bees and about 12,000 species of wasps.

In addition, wasps play an important role in the landscape ecosystem by preying on insects and eating small caterpillars and larvae that feed on food crops. *Note:* About 100 species of orchid rely almost entirely on wasps for their pollination.

Credit: GStibolt

Figure VII-8: Here, a large ground-nesting wasp called a "cicada killer" made a nest at the edge of a lawn space. Many important insect predators need un-poisoned landscapes with some soil that is not planted or mulched.

As gardeners, we can help protect pollinators from neonicotinoids by:

- Avoiding the use of these insecticides.
- Asking local nurseries or garden centers if plants have been treated with neonicotinoids, and not purchasing those that have been treated.
- Speaking out in our communities to encourage property managers to stop using neonicotinoid-treated plants in public parks, schools, municipal properties, and conservation areas.

Last Thoughts

Pollinator gardens play an important role in the neighborhood-wide ecosystem. And remember: pollinator-supporting plants can also be distributed widely across entire landscapes, not just in specialized gardens. As entomologist Doug Tallamy has pointed out, even small yards can make a big difference when it comes to supporting insects, which in turn support birds and other wildlife. If we can convince our neighbors, local schools, and churches to fill their green spaces with plants that support pollinators, then we will have helped create a bit of a wildlife corridor right outside our doors.

Additional Resources

- The USDA Forest Service has some good advice in "Gardening with Pollinators" along with a good selection of references: fs.fed.us

When we present insects from Pennsylvania, for example, with plants that evolved on another continent, chances are those insects will be unable to eat them. We used to think this was good. Kill all insects before they eat our plants! But a plant that cannot pass on the energy it has harnessed cannot fulfill its role in the food web. We plant Kousa dogwood, a species from China that supports only a few insect herbivores, instead of our native flowering dogwood (*Cornus florida*), which supports 117 species of moths and butterflies alone. My research has shown that alien ornamentals support 29 times fewer animals than do native ornamentals.

—Doug Tallamy, "A Call for Backyard Biodiversity," *American Forests*, Autumn 2009

- pollinator.org provides 32 regional 24-page brochures with general information and a good plant list of pollinator plants.
- butterfliesandmoths.org provides photos, distribution maps, and lists of larval plants for North American butterflies and moths.
- Bug Guide's "Native Bees of North America," is a good resource: bugguide. net/node
- The Xerces Society (xerces.org) works to preserve invertebrate wildlife and provides substantial resources on pollinators.

ACTION TOPIC (VII-3) Control Invasive Plants

Why This Matters

Invasive plants are exotic or non-native species that have displaced native plants in their own ecosystems and consequently damaged (simplified and/or depauperated) those natural habitats. An invasive plant can take over a new area because the insects, diseases, and foraging animals with which it co-evolved, and which kept its growth in check in its native range, are missing from its new habitat. This gives the invasive species an ecological advantage over native species.

Before labeling a plant invasive, regional nonprofit organizations study natural habitats to determine the extent of harm done by that species. The term *invasive* is not assigned lightly.

Many native plants can be aggressive in urban/suburban landscapes, usually because of disturbed soils, but by definition *native species* are not invasive (although they might be labeled *noxious*). While most exotic plants are not invasive, any aggressive exotic could become so, given enough time.

Invasive species are problematic because they:

- Outcompete, weaken, or kill native plants that are needed to support local wildlife.
- Often do not supply the kinds of habitat services—food, cover, nesting sites—that native plants provide.
- Disrupt both natural and urban ecosystems.
- Divert millions of public and private dollars for their control.

In addition, large-scale invasive plant removal efforts may include the use of power tools and large machinery, which have their own climate footprint. Disruption of the soil during these efforts releases sequestered carbon as CO_2 into the atmosphere. And the use of herbicides may damage remaining plants and animals as well as complicating the needed ecosystem restoration after the removal is completed.

Credit: SReed

Actions

❯ **Learn to identify local invasive plants (and animals).** Each region has its own set of invasive species. What's invasive in Florida will not be the same as in Massachusetts or British Columbia. There are many resources for plant identification including books, extension agents with their Master Gardeners, social media, and native plant association or invasive plant councils.

It is also helpful to know what plants are growing on your property or on community lands so that plans can be made to remove invasive species as soon as possible and to encourage the native plants, which will help make that property an important part of the effort to protect wildlife and pollinators.

❯ **Work to eliminate invasive plants from your landscape.** Remove the plants in question by using the most sustainable methods. If it's a tree, girdling the trunk (removing a three-inch strip of bark around the total circumference near the bottom of the trunk) can be a good option, because this requires no fossil fuels or toxic chemicals, and the remaining snag will become habitat for birds and more. (Be careful about doing this with trees growing near a building.) For shrubs and herbaceous plants, the solution is likely to be more complex. We recommend consulting with a local invasive plant council or natural resources department to get specific instructions and guidance.

Any plant material that is removed from a property should be properly disposed of, which usually means either burning or bagging it for the landfill.

Figure VII-9: Hardy kiwi is a good example of a non-native plant originally used for its ornamental value, with no awareness of its invasive potential. Now often planted for its fruit, in some regions of the Northeast and Midwest it has spread far beyond the gardens where it started, overwhelming some once-healthy ecosystems.

Credit: GStibolt

Figure VII-10: Manually removing invasive plants reduces soil disturbance, but the remediated area will need to be checked for re-growth on a regular basis for at least a year. This edge along a line of trees had been overtaken by two Florida invasives: wedelia and wild taro.

Plant materials from weedy herbaceous invasives may harbor huge numbers of seeds or viable runners that could start whole new populations if they just go to a municipal yard waste facility or are added to backyard compost piles or bins, which do not generate enough sustained heat to destroy seeds.

Some people bring in goats, which can be hired for a time period, to get rid of invasive plants, but goats will eat anything and everything, so some controls are necessary. Goats do solve the problem of disposing of the plant materials, though.

● **Participate in workdays** to help remove invasive plants from parks and other public wildlands. Volunteering for invasives workdays offers great opportunities to make a real difference in the local ecosystem; volunteering also provides good learning-by-doing education on the local flora and fauna. In addition, fellow participants may open doors to new educational opportunities or to new sources of locally produced native plants.

● **When you see invasive plants for sale, say something.** Despite their damage to local ecosystems and the multi-millions of dollars spent on their eradication, many known invasive plants can still be found for sale in garden centers and big box stores. For the most part, it is legal to sell them. However, unsuspecting homeowners and commercial contractors may not have any idea of the potential for harm, especially because these plants have been so widely accepted and used as landscape plants for many decades. When informed of invasive plants' capacity to displace valued native species, most people choose to avoid buying or using them.

Important caveat: There are claims that some cultivars of invasive plants are sterile and therefore harmless. This has been proven to be mostly false, as these cultivars eventually change into fertile forms that can interbreed with the native population. We need to completely abandon using any plants that have proved to be invasive. In addition, we should be looking into ways to identify and discontinue using any new plants that show likely potential for invading our natural areas.

Last Thoughts

It is commonly understood that invasive species cause problems for the environment by out-competing less aggressive species and thereby disrupting the working balance of relationships within established ecosystems. From the standpoint of climate change, invasive plants matter in several other ways,

One Case History

Kudzu is native to Japan and southeast China. It was first introduced to the United States during the Philadelphia Centennial Exposition in 1876. Then, in the 1930s through the 1950s, the US Soil Conservation Service promoted it as a great control for soil erosion, and it was widely planted in the South. Little did they know that kudzu someday would be overtaking and growing over anything in its path. The vine that ate the South!

Credit: GStibolt

Figure VII-11: Kudzu was widely planted for erosion control but it has taken over whole native forests. This vine kills not only native trees and shrubs, but also all the wildlife that depends upon those ecosystems.

I cannot stress enough the menace of invasive species. Some authors, thankfully few in number, have naively suggested that in time alien plants and animals will form "novel ecosystems" that replace the natural ecosystems wiped out by us and our hitchhiker companion species. There is evidence that some alien species of plants "naturalize" in island environments, in other words genetically adapt to them by natural selection. But this occurs only where the diversity of plant species is low and offers a relative abundance of empty niches for alien species to fill.

— E. O. Wilson, *Half-Earth*, 2016

too. First, every region needs stable, resilient ecosystems to keep ecosystem services working smoothly, and this is especially important as climate change stresses all ecosystems. And second, the cost of controlling and removing invasive species can be huge. This is true not only in terms of money spent, but also in the consumption of fossil fuels (getting crews to and from job sites, running equipment and tools), the application of herbicides (with their likely toxic effects) and, an often overlooked consequence, the emission of CO_2 in all of these activities.

ACTION TOPIC (VII-4) **Design Perennial Gardens to Serve Local Ecosystems**

Why This Matters

A hundred fifty years ago, the European upper classes were expected to have extensive perennial and annual gardens filled with mostly exotic plants trimmed to a fare-thee-well. Even present-day garden magazines and TV shows set impossibly high standards for luscious border gardens; photos and videos show everything in full bloom and perfectly trimmed. The reality of dealing with living plants is that it is an ongoing process, not just a snapshot in time; a more natural result can be a beautiful array of plants decorated with the birds and butterflies that find food and refuge there.

Perennial gardens and borders provide important habitat for a wide variety of wildlife in transition areas at the edges (*ecotones*) of groves or wooded areas. They can be used as foundation plantings instead of shrubs that outgrow the space and need constant trimming or periodic replacement. In addition, herbaceous gardens can serve as butterfly gardens and include a variety of edibles.

Credit: GStibolt

Figure VII-12: A lawn mower can easily navigate the edge of this wild area. Later, as the plants spread, the border can be moved further into the lawn area.

Actions

● **Design the size and shape of the garden or border for ease of access.** Keep beds narrow so nothing is out of reach, or build narrow paths in the beds between the planting areas or next to fences or buildings.

If the garden abuts a lawn, design the edge next to the lawn so that there are gentle curves and no vertical dividers between the lawn and the bed. The goal of this interface is to allow the mower to make one pass along that edge with no need for further power tools. See Action Topic I-2 on page 23 for more information on reshaping lawns.

● **Choose the plants.** As discussed in VII-1, choose the right plants for the location. If the border is in a shady location, such as the north side of a wooded area, one or two spreading fern species can work very well as an understated border. If the garden is a wide swath in a sunny area that's highly visible, a large number of plant types may work well for a more exuberant space. Two or three bunching grasses planted in a grouping will give the space structure, but then you can add a variety of forbs that bloom in different seasons. If there are low spots, select species that thrive in that microclimate. (*Forbs* are herbaceous flowering plants that are not grass-like.)

Credit: SReed

Figure VII-13: This segment of the High Line in New York City demonstrates how grasses, flowers, and shrubs can all combine to create a vibrant ecosystem.

Credit: GStibolt

Figure VII-14: Coneflowers at Lurie Garden in Chicago are allowed to go to seed, and look how beautiful the results are. Even more important, the seedheads provide essential food for many birds in the fall.

When purchasing plants, look for smaller, younger specimens; they will save you time and money in two ways: (1) they will adjust more quickly to the garden, so less irrigation is needed; and (2) they are more likely to survive, so you won't need to replace them. In addition, younger plants are also more climate-wise, because the grower can produce them more quickly, which uses less water and fewer resources.

◗ **Avoid deadheading in a native landscape.** Traditional gardening guidelines say to cut back the dead flower heads to make them look neater and make some plants bloom again. Humans and pollinators enjoy the extra flowering cycle, but maybe the seed-eating birds needed that seed for their young, or maybe the extra flowering took a toll on the plants (after all, if nature designed the plants to have a fallow year in between flowering years, maybe there's a reason!). For a lower-maintenance yard, just let the plants follow their natural cycles. See Action Topic VII-2 on page 205 for information on pollinator gardens.

Last Thoughts

Even though perennial garden borders are planted mainly to create something special and beautiful, each of these gardens is a working ecosystem that is constantly changing, just like every other part of our landscapes. Plants are not isolated entities but participants in a system constantly in flux with the other inhabitants—both plants and animals. So even a carefully designed border garden will change over time—some plants will die, while others may grow well for decades. This is something we should learn to welcome. New plants and combinations can be added along the way to add variety and new beauty and attract a new set of pollinators.

Gardens also teach the necessary if un-American lesson that nature and culture can be compromised, that there might be some middle ground between the lawn and the forest, between those who would complete the conquest of the planet in the name of progress, and those who believe it's time we abdicated our rule and left the earth in the care of more innocent species. The garden suggests there might be a place where we can meet nature halfway.

—Michael Pollan, *Second Nature: A Gardener's Education*, 1991

VIII

URBAN ISSUES

Overview

Urban Heat Islands

Of the many complex problems cities will confront in the coming years, urban *heat* is likely to be one of the most serious. Here, we focus on landscape-related actions we can take to help address some of the predicted extremes.

Cities are frequently hotter than their rural and suburban surroundings. This results from an unfortunate combination of too much heat absorbing concrete, asphalt, and steel...plus too many heat-generating industries, vehicles, and air conditioners...plus too little heat-reducing vegetation...plus crowded buildings that prevent nighttime breezes from carrying away the daytime heat.

Together these conditions produce a phenomenon called the *urban heat island*, which involves these detrimental conditions:

- Higher temperatures bring increased use of air conditioning and fans. This raises electricity demand, which increases emissions of pollutants and greenhouse gases at the power plant, exacerbating climate change. Plus, as they run, air conditioners themselves produce heat, which gets dumped into the mix of urban air.

- City heat can also lead to *temperature inversions*, trapping pollution and ozone in the lower air layers, where we live. This can cause respiratory problems, especially in children and seniors; in addition to being a serious health concern, this can also increase added energy demand (doctor visits, hospitals, etc.).

- Hot pavement can require more frequent maintenance.
- Finally, stormwater that runs off hot roofs and pavement carries that heat into storm sewers. This ultimately raises the water temperature in streams and rivers, changing habitat conditions and harming many species. A 2003 EPA study showed that when the temperature of urban pavement in Chicago was 20–35 degrees higher than the air, rain runoff from that pavement was about 20–30 degrees hotter than runoff in nearby rural areas.[1]

In This Section

In addition to large-scale governmental initiatives to help reduce the buildup of city heat (such as by burying roads or converting ex-industrial semi-wasteland into green spaces), there are many small-scale landscape-related steps we all can take to contribute to the solution. These steps are detailed in the following *Action Topics*:

- VIII-1: Maximize Urban Vegetation
- VIII-2: Create More Green Roofs and Green Walls
- VIII-3: Design Cool Roofs
- VIII-4: Support the Use of Cool Pavement Techniques

Just How Bad Is Urban Heat?

- In a 2014 study, the nonprofit organization Climate Central found that in the past ten years, of the 60 largest cities in the contiguous US, the largest single-day temperature difference between city and country was 27°F. And the average *single-day* differential across all 60 cities was 17.5°F.
- For the *entire summer*, while the city-country heat differential averaged 2.4 degrees for all 60 cities (with nights being on average 4 degrees higher), many cities were dramatically hotter.

In Las Vegas, the whole summer differential was 7.3 degrees, in Albuquerque, it was 5.9 degrees. Even many northern cities like Denver, Portland, and Minneapolis were almost 5 degrees hotter than their outlying towns.

- Further, nearly half of the cities studied have been hotter every year since 1970. And although summer heat has been rising steadily across all of North America, the study showed that temperatures in 45 of the 60 cities rose even more than in rural areas.[2]

Two other topics that relate to urban landscaping include:
- VIII-5: Choose Climate-Wise Lighting
- VIII-6: Create Habitat Corridors

ACTION TOPIC (VIII-1) **Maximize Urban Vegetation** »

Why This Matters

All plants cool the environment in two ways:
- By shading the ground and nearby buildings.
- Through evaporative cooling, also called *transpiration* (see explanation in "A Primer on Water Chemistry and Plants" in Section III on page 75).
- Although trees provide the most noticeable and extensive shade, keep in mind that shrubs, vines, perennials, and grasses also shade the ground and contribute substantially to the evaporative cooling effect.

Here we focus mainly on trees because they are simultaneously the most challenging, important, visible, and studied of the vegetation we need in cities. However, many of these suggestions apply just as well to shrubs and vines. Bottom line: the more plants, the better!

Actions

▶ **Plant shade trees in every possible spot.** Trees located south or southwest of a building offer the most shading value, but even trees east of a building will help by reducing air conditioning demand as the day begins to heat up.

▶ **Where possible, plant trees in groups** so their shared root zone (and shared mulch area) can be as large as possible so tree roots can support/communicate with each other for maximum health. Where space is limited, choose

> ### How Much do Trees Cool a City?
> - Trees can keep 70–90% of the sun's energy from reaching the ground.[3]
> - Walls and roofs can be cooled 20–45°F by shade trees.[4]
> - Tree shading can reduce temperatures inside parked cars by 45°F.[5]
> - Evapotranspiration can reduce peak summer temperatures by 2–9°F.[6]

narrow trees and trees that have evolved as *understory* trees, i.e., smaller stature trees that can thrive in the shade below larger canopy trees.

⊙ **Plant mostly deciduous trees,** because they produce shade in summer but allow winter warmth to penetrate into windows and onto pavement.

⊙ **Take good care of trees.** In summer, when trees are most stressed, they should be watered, even more so than lawn, especially if they've been planted in the last couple of years. Prune broken limbs to help trees heal. (But check with the local public works department or tree warden before watering or pruning any street trees.) See II-1 on page 52 for more about taking care of trees.

⊙ **Avoid "planned obsolescence."** Many existing street trees are over 100 years old, so we know it's possible for urban trees to survive longer than the current average of seven years for newly planted trees. Many new policies have been adopted to improve the way street trees are planted—adjusting the size and

Trees and Drought

When soil gets dry, lawn grass and perennials might turn brown, and this draws our attention to their condition. But for trees, the effects of drought often starts with dieback of the highest branch tips, which we are least likely to notice. Lawns survive drought by going dormant and popping back to life when water returns later. Perennials are easily replaced. Trees and large shrubs, on the other hand, might not ever fully recover from prolonged drought. If they die, their absence will be a significant loss of cooling shade and transpiration, habitat, time, and money.

Credit: SReed

Figure VIII-1: That's the spirit: put plants in every possible square inch!

depth of tree pits, for example. For planting urban trees on private property, check with natural resources agencies, a public works department, or local nature center for advice about trees most tolerant of urban conditions in your region. You can also ask them to recommend planting methods.

❯ **Encourage Public Works, Planning Commissions, Park Districts, and City Tree Departments** to plant more trees at every opportunity, and to publicize these efforts so the nearby residents will get involved and take pride in the new vegetation. Also encourage Public Works to add curb cuts and to build street-side rain gardens to reduce the stress on storm drainage systems during severe weather. They could be sized to accommodate multiple trees, which can play a big part in moving the water from the streets into the air; and with this extra water, the trees survive longer.

Figure VIII-2: This flowering dogwood tree, which might typically thrive in the understory of a mature oak forest, seems perfectly happy to grow in a generous patch of shady Brooklyn soil.

Credit: SReed

❯ **Plant trees specifically to shade parking lots.** If there's not enough space for a tree, plant shrubs or other vegetation to at least cover the ground. Shade helps extend the life of the pavement and helps cars run more efficiently.

Last Thoughts

Trees are good for cities. In addition to reducing urban heat, they also:

- Improve air quality, reducing ozone and particulates from car exhaust.
- Remove pollutants from the air, and improve overall human health.
- Buffer noise.
- Help reduce stormwater runoff by siphoning water up via their roots to branches and foliage, where it is transpired into the air (and where it then cycles back as rain).

Credit: SReed

Figure VIII-3: These mature and stately sycamore trees are prized for their shade and cooling; this park's porous pavement helps nourish and water their roots, enabling the trees to reach a healthy old age.

- Help to speed up wind, which helps to reduce urban heat.
- Provide invaluable habitat to countless non-human organisms, who in turn make the world go 'round.
- Store carbon in the trunks and roots.
- Are beautiful, and their presence in our lives makes us happy.

For all of these reasons, one of the most important things all of us can do to help curb climate change and improve our own lives, is to plant trees, shrubs, and whatever leafy vegetation can be supported by existing rainfall and a modest amount of human assistance.

Figure VIII-4: A good way to support urban vegetation is to encourage municipal authorities in charge of infrastructure to build more modern tree pits, which (unlike this one) will enable the soil to absorb more water and give roots more space to grow.

Credit: SReed

Whether we and our politicians know it or not, Nature is party to all our decisions, and she has more votes, a longer memory, and a sterner sense of justice than we do.

— Wendell Berry, commentary on jacket of
The Dying of the Trees, by Charles E. Little, 1997

The Stresses on Urban Trees

Most trees can thrive in the nearly optimal conditions of consistently moist but not soggy soil, an approximately neutral soil pH, and abundant sunshine. Unfortunately, trees planted in cities rarely experience any of these ideals; instead, they are frequently subjected to a wide array of extreme stresses not experienced by trees growing in forests and rural or suburban landscapes. For any urban tree-planting project, plan ahead to address these challenges:

- *Poor/compacted soil*: The major problem for city trees is soil that will not allow the roots to absorb water, oxygen, and nutrients. Tree pits are often too small, and the soil poor and/or too compacted, to support root growth.
- *Inadequate soil moisture*: Although many trees do adjust to less-than-optimal soil moisture, it's important to help newly planted trees by watering them regularly for the first few years.
- *Extremes of sun and shade*: Most trees are adapted to grow best in full or nearly full sun, but cities contain many pockets of deep shade cast by tall buildings. Make sure to consider sun and shade when choosing what and where to plant trees.
- *Salt in the soil*: Road salt is especially harmful for trees if it is used during an early snowfall in October or November, or if it is spread in the late spring, because at these times tree roots are actively pulling moisture up into the trunk and leaves. In salted areas, choose species with a demonstrated tolerance for salt uptake.
- *Soil pH*: Most city soil has an alkaline pH due to the limestone in concrete so often used in cities. Any trees that require acid soil will require constant amendments, so it's best to avoid these.
- *Insects and disease*: Trees under stress are more vulnerable to insect attack and diseases. The best way to protect trees from these problems is to plant them in good soil, and then keep them well watered and mulched.

ACTION TOPIC (VIII-2) Create More Green Roofs and Green Walls

Green Roofs

A green roof is a layer of vegetation growing in a medium that covers some or all of a roof. This blocks sunlight from being absorbed by the roof's surface, which in turn reduces the amount of heat that then flows through the roof into the building. A cooler building demands less air conditioning. And in winter, the extra layer of growing soil can help insulate the building. *Note*: Green roofs should never be used in place of insulation.

In addition to reducing cooling and heating costs for building owners, green roofs also provide several public benefits.

- Reduced electricity usage will (until all electricity is produced renewably) reduce greenhouse gas emissions at the power plant.
- Green roofs provide habitat for urban wildlife, especially for pollinators, many species of which are essential for urban farming.
- They can support urban food gardens.
- They can be designed to be part of outdoor spaces where people can relax and socialize away from the busyness of the street level.
- Larger plants can help suppress city noise for a more peaceful environment.
- Some roofs help manage urban stormwater runoff: the plants themselves take up and transpire the moisture, and the porous soil medium also temporarily stores excess water, delaying its release, which reduces pressure on storm sewer systems and infrastructure. Some green roofs include the ability to sequester stormwater so that it can be used for irrigation during dry times.

One of green roofs' greatest assets, though, is their ability—when enough of them exist—to cool the air all across a city. One study performed in Toronto, Canada, concluded that adding green roofs to half of the available downtown buildings could cool the whole city by up to 1.4°F. Further, if these roofs were irrigated, the resulting increased transpiration could reduce temperatures by as much 3.5°F.[7]

As a result of all these benefits, green roofs can play a significant role in reducing urban heat in the decades ahead. The organization Green Roofs for Healthy Cities has recorded an annual increase in green roof installation across North America of 18.5% in 2015, with Washington DC ranking first, and Toronto second. And, increasingly, many cities are establishing policies and programs that both reward and support investment in green roof projects.[8]

Although each green roof is unique in appearance and structure, all are constructed as either one of two basic types: *extensive* or *intensive*. Here's a summary of the main characteristics of each type of green roof:

Credit: SReed

Figure VIII-3. This extensive green roof at the University of Massachusetts in Amherst cools the building below and helps store/manage stormwater.

Table 1: Roof type comparison

Extensive	Intensive
Shallow (4–6″ deep) soil or growing medium	Soil can be deep as conventional garden soil
Grow only non-woody plants, selected for their tolerance of extreme conditions	No limit on what can grow, including even shrubs and large trees
Generally not irrigated	Generally irrigated
Low or minimal maintenance	Intensively maintained
May not require extra structural support	Definitely needs extra structural support
Built on flat or pitched roofs (up to 7:12)	Built only on flat roofs
Intended mainly for insulation. Less effective at reducing stormwater runoff. After installation, there must be a minimum of foot traffic.	Intended for insulation, stormwater control, and possibly public use/enjoyment. Increasingly, green roofs are used to grow food to provide local food sources in so called food deserts.

Credit: GStibolt

Figure VIII-6: Small-scale green roofs like this one in Seattle, Washington, make the most of regional rainfall, absorb some amount of stormwater, and add diversity to a landscape.

Credit: © zgurski980/Adobe Stock

Figure VIII-7: Green walls—which contain an attached growing structure and dedicated irrigation method—are a popular new idea for cooling buildings and adding vibrant vegetation to a streetscape.

The choice of which type will best meet your needs depends upon several interrelated factors: the purpose the green roof is intended to serve; the desired look/appearance; what sort of roof structure exists or will be built; how much irrigation and maintenance will be provided; and anticipated budget.

Note: For green roofs to substantially reduce stormwater runoff, they need to be *intensive*. And this kind of roof structure must be heavily engineered to support the load of the added depth required to accomplish this goal. This can be done relatively easily with new buildings and municipal projects. At the residential scale, however, roofs are not generally strong enough to support the depth of soil needed for an intensive roof.

Green Walls

Like green roofs, green walls can cool a building and reduce utility costs, either by reducing the need for air conditioning altogether or by shading air conditioners, making them more efficient. Also known as *living walls* and *vertical gardens*, green walls are partly or totally covered with vegetation. They consist of a growing medium held in a structural framework that is attached to the building, and (sometimes) an integrated system for delivering water

A Good New Idea?

Vertical forests are urban buildings wrapped by nearly continuous balconies that have been built on every floor of the building to support large and extensive vegetation that will blanket the exterior. Like green roofs, this vegetation will soak up urban air pollution, produce oxygen, cool the air and enhance local biodiversity. Although these structures are mainly an architectural/engineering concept, their success will depend on users and residents understanding which vegetation to plant and how best to support its growth in challenging conditions such as hot southern and western exposures, cold and shady northern exposures, swirling winds at the lower levels and high-speed winds at the upper levels, and the need for regular irrigation. Nonetheless, the idea of vertical forests holds great potential for reducing urban heat and helping wildlife.

to the plants. *Note*: In contrast to green walls, green *facades* consist of plants climbing a wall from a soil source (a container or the ground) at the base of the wall.

Green walls can also fall into the same two categories as green roofs. Extensive green walls will be mainly self-tending after the installation of the substrate and the hardy plants, which can survive on the local rainfall and droughts. Intensive green walls are often used to grow edibles and usually use hydroponic irrigation systems.

Green walls may be built inside a building or as an exterior wall. They may be freestanding, attached to an existing wall, or integral to the structure of the building. Like green roofs, green walls help cool a building and reduce demand for air conditioning. They also can help reduce urban heat, support pollinators, and contribute to urban biodiversity.

Last Thoughts

As scores of historic ivy-clad colleges across the country attest, the cooling effect of vegetation on buildings is nothing new. What is new, however, is the idea that we can—and should—design and construct buildings specifically to achieve this goal. New policies and technologies support this intention, and

> Green roofs, climber-clad buildings, and rain gardens need to be integrated with passive air-conditioning systems, sustainable waste management, and sustainable energy capture and other technologies, so that they all work together. Isolated green roofs can only do so much, but integrated with other technologies there is a powerful synergy that will enable us to share an increasingly crowded planet.
>
> —Nigel Dunnett and Noel Kingsbury,
> *Planting Green Roofs and Living Walls*, 2008

now the green roof and green wall markets are growing like weeds. According to the website Green Roofs for Healthy Cities, new green roof construction in the top five metropolitan North American cities reached over 3.5 million square feet.[9] Statistics for green walls are harder to track, but the trend is steadily upward.

ACTION TOPIC (VIII-3) **Design Cool Roofs** »

Why This Matters

Cool roofs are designed to absorb less heat than conventional roofs. This mainly involves making them more reflective, through the use of lighter colors. However, new technologies also aim to produce surfacing materials with higher *thermal emittance*, or the ability to give off absorbed heat more quickly. The ultimate goal is to reduce surface temperatures and hold onto less heat during the day, which means less heat radiated from the surface at night.

Conventional shingle, asphalt, and tile roofs can be 55–85°F hotter than the air, while cool roofs tend to be only 10–20°F hotter. And while traditional roof surfaces may absorb 85–95% of the energy that reaches them, the coolest cool roof materials can reduce that rate to 35%, substantially lowering the amount of heat that is transmitted into a building.[10]

Actions

Both low-sloped and steep-sloped roofs can benefit from cool roofing techniques, but each type needs a different approach.

Low-sloped roofs may be flat or have a maximum pitch of 2 inches of vertical rise in 12 inches of run (2:12). These roofs are most commonly found on commercial and industrial buildings and sometimes on apartment buildings,

Credit: SReed

Figure VIII-8: Many new buildings are being designed with roof gardens and white/reflective structures that together lower utility costs in the building and help reduce urban heat.

but rarely on homes. In the past, this type of roof would often be waterproofed with thick liquid asphalt that was either painted or sprayed on.

Two types of cool roof materials can be used to replace this asphalt. *elasto meric coatings* that are sprayed on, or *flexible sheeting* that is laid on the roof with seams that are glued, taped, or heat-sealed. Both are manufactured to have both high reflectivity and high thermal emittance. Most cool roof programs tend to focus on the low-sloped sector of the roofing world.

Steep-sloped roofing alternatives, however, are becoming more available every year. Instead of the traditional dark-colored clay or concrete tiles, with a reflectance of 10–30%, manufacturers are now producing tiles with a special infrared-reflecting pigment that raises that figure to 25–70% reflectance.

Asphalt shingles are currently used on more buildings than any other roofing material, partly due to their low initial cost. Now, manufacturers are working to raise the shingles' reflectivity to as much as 65%, mainly by using lighter colors and infrared-reflecting pigments. And the popularity of light-colored metal roofs is also growing steadily, for the same reason.

The ultimate in cool roofing is the emerging field of photovoltaic (or solar) shingles. Beyond being highly reflective and generating electricity for the building, they can now actually take the place of conventional shingles, tiles, or shakes. And, unlike standard roof-mounted solar PV arrays, which require rack-mounting that penetrates the roofing material, solar shingles are attached directly to the roof sheathing. As of this writing, solar PV shingles are an emerging technology with a lot of potential, but they are not yet widely available. (See VI-8 on page 186 for more about incorporating renewable energy sources in landscapes.)

Last thoughts

Cool roofs can save on utility costs in several ways: through reduced energy demand, smaller AC equipment, longer roof lifetime, and rebates and incentives. When building a new roof, the installation costs can be the same as a regular roof, but converting a standard roof might cost more than the expected savings. "Cool roof calculators" found on the internet can help with figuring this out.

Keep in mind that cool roofs tend to have the greatest benefit in hot climates. In cool climates, by comparison, they have the potential to *increase* energy costs by reducing wintertime heat gains. As with many of the guidelines in this book, the right decision will require some homework. And local expertise will also help.

ACTION TOPIC (VIII-4) **Support the Use of Cool Pavement Techniques**

Why This Matters

In many metropolitan areas, pavement covers more land than do buildings and vegetation.[11] Conventional, impervious paving materials like concrete and asphalt can absorb 60–95% of the sunlight that reaches them, and their surface temperatures can get as high as 120–150°F.[12] This daytime heat transfers into the mass of pavement below, where it is held and then released at night. Further, rainwater running off hot pavement carries that heat into streams and waterways, altering habitat, which harms both plant and animal species. Of course, most urban pavement is a municipal responsibility, so most of the actions we can take as individuals amount to encouraging and supporting the installation of cooler pavements.

Actions

As with cool roofs, the solutions to this set of problems involve lowering surface temperatures, both by providing more shade and by using more reflective materials. Unlike roofs, though, which only need to waterproof a building, pavement must meet many requirements and serve many purposes. The surface must be strong enough to withstand heavy and extended traffic. And paved surfaces can range from huge interstates to tiny front paths, with a vast variety of functions and requirements in between.

Pavement, more than roofs, gets dirty and worn, which can change its reflectivity. Over time, concrete darkens and asphalt lightens. And it is difficult to measure the effect of pavement heat because the movement of vehicles creates convection currents that move heat around.

For the purpose of shading and cooling pavement, we have described actions to increase urban vegetation in VIII-1. Here, we address two other methods for decreasing pavement heat. The first involves increasing reflectivity; the second involves making surfaces porous so rainwater can seep in rather than running off—which turns out to be a surprisingly effective heat-reducer.

◉ **Steps to increase solar reflectance include:**

- When appropriate for a project, use concrete, which is naturally light-colored, instead of dark asphalt.
- When paving with asphalt, choose light-colored aggregate (crushed stone binding materials).
- On low-traffic areas, use resin-based pavements, or colored asphalt or concrete.
- Use pavement surfacing applications, or *white-topping*, that increase reflectivity.

◉ **Permeable pavement** (also called *porous* or *pervious* pavement), which originated as a way to manage stormwater, has been discovered to also provide a significant cooling effect, especially when wet. Typically used on lower traffic areas, permeable pavement contains voids that allow water to drain through into belowground storage chambers and the surrounding soil. The pavement may

Reflected Heat

Cool pavements stay cool by reflecting more of the sun's rays than conventional pavements do. Although many of these reflected rays bounce back into space, a portion of them rebound into the windows of nearby buildings. This can increase the amount of sunshine in those buildings, potentially reducing lighting costs. However, depending on the age and design of the buildings, this can also cause these buildings to heat up, which increases demand for air conditioning. So it's a balancing act.

Credit: SReed

Figure VIII-9: Light-colored pavement and porous surfaces together reduce solar absorption and absorb rainwater to cool the ground.

Credit: SReed

Figure VIII-10: Concrete/grass pavers create a firm, porous driving surface, ideal for low-traffic areas. An alternate, low-carbon solution is to use the same pavers but leave the cells empty or fill them with crushed pebbles, to eliminate the need for added topsoil and future mowing.

be made of asphalt, concrete, brick, or block filled with gravel or rocks, which may themselves also be light in color. In some situations, the permeable pavement may be vegetated. This can involve using grass pavers or concrete grid pavers arranged in a lattice and filled in with grass or other low-growing plants. (For more, see "What about Porous/Pervious Pavement?" in X-4 on page 287.) In a low traffic driveway, only the tire tracks might be paved, with grass or other vegetation occupying the rest of the width.

Note: Pervious concrete and asphalt will gradually become less effective as the voids fill up with silt and grit. Vacuum trucks and/or power washers may be needed to clear the voids. For this reason,

What about Solar Roads?

As of this writing, the idea of replacing pavement with solar panels is being tested in several countries. This involves using plastics and crushed glass to produce durable, anti-slip surfaces laid atop ordinary solar panels, creating a road capable of handling even 18-wheelers. The main advantage here is that whereas solar farms often occupy land that could better be used for agriculture, parks, or wild areas, existing roads are available without clearing additional land. Further, the electricity generated by a solar road might even power the vehicles travelling on it. Major challenges, for now, include cost and, perhaps toughest of all, making the surface able to withstand a snowplow.[13]

porous paving is a complex solution best suited to large projects with budgets that include maintenance.

Last Thoughts

As home- and/or property-owners, we can choose from these cool pavement techniques for surfacing driveways, patios, and paths. But even if we don't own property, we can still actively encourage public agencies and political bodies to choose cool paving for all public projects.

ACTION TOPIC (VIII-5) **Choose Climate-Wise Lighting** »

Why This Matters

It is possible to light our landscapes as efficiently as possible and do it in ways that have the least impact on nighttime wildlife (and the neighbors!). Not every landscape needs to have outdoor lights, and skipping them altogether will save the most energy. But if you do choose to install outdoor lights, consider taking some or all of the low-carbon actions below.

Actions

▶ **Use the right energy source for the job.** Line voltage (usually 120 volt) is supplied by the power grid to our houses. This strong source of electricity is appropriate for security lighting and other high-wattage fixtures. When used for outdoor lighting, however, all wires must either be buried 24″ below ground or placed in a conduit; both solutions require either advance planning during construction or the disruption of an established landscape. With line voltage, it's all too easy to choose fixtures that

> ### Night Light
> Nighttime lighting can affect the blooming cycles of the many plants that use day length to determine when to bloom. It can affect the darkness needed for many species of moths and their need for night-blooming plants. Keep in mind that while we don't notice most of the moths around us, there are actually 15 times more species of moths than butterflies in the world, and many of them play important roles in our local ecosystems as pollinators and as a food sources for larger animals.

provide more light than necessary, potentially wasting electricity and increasing the impact on wildlife and the night sky.

For much residential lighting, low-voltage and solar electricity is sufficient and a wiser choice. Low-voltage systems convert 120-volt to 12-volt power via a small transformer plugged into an existing outside receptacle. This electricity supply is easy to install by non-professionals. When the wires are buried in a trench that's at least 6″ deep, they need no conduit and are completely safe. Solar-powered lighting generally consists of a small solar collector placed near or in a fixture, with a rechargeable battery (also part of the fixture) that stores daytime electricity for use at night. These systems are easy to set up and move. Their main disadvantage is the same as with all solar power: they require 6–8 hours of sunlight to charge to full capacity. Solar lighting is most suitable in situations that don't require full or bright lighting all night long.

❯ **Arrange electric circuits to allow maximum control.** The best way to save lighting energy and minimize light pollution is to make sure we turn lights on only when they're needed, and turn them off when they're not. And we should pay attention to what it is we are providing the light for. Lighting your front porch shouldn't also mean illuminating the whole driveway. Using your patio, deck, or pool shouldn't necessarily involve lighting the garage or back porch. The best way to address this is to make sure light switch circuits are set up to allow for individual control and are within convenient reach.

❯ **Install controls to provide "just-in-time lighting."** Three kinds of specialized lighting controls enable us to prevent or reduce wasted energy on our outdoor lighting systems.
- Motion detectors enable a portion of the landscape—the driveway, entry path, sidewalk, porch, etc.—to be lit only when someone is approaching, then turning off and leaving that space dark the rest of the time.
- Photo-sensors save energy by turning lights on and off in response to changes in ambient light conditions. This prevents lights from staying on all day when they're not needed. They are especially useful in places where no one is likely to be present to switch lights on or off manually.
- Timers, like motion detectors and photo-sensors, help us reduce unnecessary lighting. Timers are simple mechanisms that can be set to turn lights

on at dusk and off at dawn, and they're particularly useful in situations where nighttime lighting may be overlooked or forgotten, and then remain on during the day. Timers that will control low-voltage outdoor lighting may be simple units that are plugged into the system's transformer. These outdoor devices may also contain photo-sensors that turn lights on at different times through the year as days lengthen and shorten, then activate the timer to turn lights off after a set amount of time.

Figure VIII-11: Low-voltage fixtures with efficient LED lights can be the most climate-wise choice for lighting in many gardens.

▶ **Use the most appropriate light bulb.** We all understand that a 100-watt bulb is brighter than a 60-watt bulb. But watts only tell us how much power is being used. A bulb's brightness, the light we can actually see, is measured in *lumens*. (A lumen is a measurement of light equal to the light of one foot-candle falling on one square foot of area. And a foot-candle is the amount of light visible one foot away from a burning candle.)

One of the best ways to save energy and money is to use bulbs that put out the most light for the least input of power. Compact fluorescent light bulbs and LEDs (light-emitting diodes) are the two most efficient choices (as of this writing).

▶ **Point lights downward, not at the sky.** Lights that shine up into the night sky cause several problems. This *sky glow* reduces our view of the night sky and diminishes our pleasure in (and perhaps biological need for) darkness. Lights pointing upward instead of downward (where the illumination is needed for safety) waste energy, increase our carbon footprint, negatively affect night-flying wildlife, and can also seriously compromise astronomical observations.

The solution lies in using fixtures that direct downward, not sideways and not upward, by tucking the

Figure VIII-12: Down-lights point the illumination where it is needed—toward the ground—and prevent light pollution that can be harmful to wildlife (and intrusive to neighbors).

bulb up inside some sort of hood or shield. When these *full cutoff* fixtures are used, no light leaks out above the horizontal plane of the cover.

Last Thoughts

Planning for just-in-time lighting will not only save energy and reduce our carbon footprint, it will also help ecosystems in the landscape be more resilient, because darkness is an important factor in a natural environment.

ACTION TOPIC **VIII-6** Create Habitat Corridors ≫

Why This Matters

In the 20th century, as Western societies began to notice environmental losses piling up, we collectively decided to set aside and protect some natural areas. Each reserve was its own little bubble, an island of refuge in an expanding sea of human development. For quite a while, we believed this was the best way to take care of the natural world. And this belief was not necessarily wrong. In fact, such conservation is still happening, and that is all to the good.

But our understanding has evolved, and we now realize that protected areas also need to interact with their surroundings. Some species need to travel long distances to mate and survive, or they need a wide range of conditions at different times of year. Some species need a variety of food or nesting materials that might not be available within their own locality. And nearly all species benefit from a large enough breeding area to preserve genetic diversity and avoid inbreeding. In fact, in undisturbed nature, creatures frequently roam beyond their immediate territories in search of these necessities.

Unfortunately, the natural world has become increasingly fragmented by human development. And this fragmentation makes creatures' natural

> Movement across the landscape is a necessity for animal life. Before man's advent these lines of movement were nicely adjusted, as animal trails, to physical needs for water, food and minerals. They were also adjusted to the character of the terrain, winding back and forth across slopes and following a path of least effort.
>
> —Paul Sears, *The Living Landscape*, 1962

behaviors dangerous, difficult, or sometimes impossible. Some species are
highly mobile while others can't go far,
and still others don't move at all. Plants
can migrate only by making seeds that
move on a breeze, in a current, or with
the assistance of other species in motion, either inside when animals have
eaten seeds, or outside when seeds have
clung to the fur, skin, or clothing.

Climate change is now further com-
pounding the problem by causing large
numbers of species to leave familiar ter-
ritory in search of more hospitable and
livable habitats. More creatures than ever before are on the move. Some will
only pass through new areas, while others will stay if conditions allow. Some
will migrate north to find cooler conditions, while others might migrate to
higher elevations.

Yet this movement, a natural adaptation to changing conditions, leads to
two basic kinds of problems. First, new species arriving in established communities can disrupt established systems of predation, competition, symbiotic associations, and other natural processes. And second, many obstacles physically
stand in the way of physical movement. Cities are large and getting larger. Like
a dam, they can sometimes entirely block north-south migration. Channelized
or dammed rivers and urbanized waterfronts; 4-, 8-, and 16-lane highways; immense sprawling suburbs mostly devoid of useful habitat—to say that we have
dramatically reduced species' options for getting around is an understatement.

What's needed now are more ways to help creatures to move as they need.
Professional conservation planners and adaptation ecologists may work to create urban greenbelts, wildlife bridges over highways, and other natural linkages. But even we non-scientist private citizens can help make a difference for
wildlife.

How? By creating lots of pathways and resting places that species might
choose to use when moving through an urban or suburban region. The goal is
to have so many of these that they add up to enough. What we're trying to do

Credit: SReed

Figure VIII-13: Cedar waxwings gobble up the early-ripening fruit of serviceberry trees. These birds are among the many migratory species vulnerable to fluctuations in flowering/fruiting times.

What Is a Habitat Corridor?

Also known as *wildlife linkages*, *dispersal corridors*, and *ecological connectivity*, habitat corridors are fragments of vegetation within a developed environment that enable wild creatures to move around. While the term *corridor* implies a narrow hallway, and while much wildlife does in fact migrate within linear pathways like hedges, riverbanks, rail and transmission line easements, these connecting links could just as well be a mosaic or series of patches, or "stepping stones." In general, the less edge there is relative to the interior of the patch, the greater number of species the patch is likely to support.

Figure VIII-14: The elevated High Line in New York City—used by people and wildlife—exemplifies the best of our efforts to provide natural linkages in urban centers.

Credit: SReed

is offer an opportunity for creatures to find what they need to survive, either when natural areas are too fragmented and far apart or when habitat has become inhospitable or unlivable.

Actions

Our individual landscapes and yards are full of opportunities to create passageways, stepping-stone habitats, and temporary or permanent refuges for wildlife on the move. Toward that end, consider taking some of the following actions.

◗ **Preserve as much natural landscape as possible,** when building on or developing vegetated land. (See IV-3 on page 113 for more.) Protecting ecosystems and their complex functioning right from the start is much easier than restoring those assets afterward.

❯ **Identify any nearby natural or conservation areas,** and try to plant in a way that expands that particular type of ecosystem by using similar plant species.

❯ **Wider and rounder patches are generally preferable to long, linear corridors.** But if you only have space for a hedgerow or a narrow band of shrubs, these are preferable to no natural areas at all.

❯ **Use mostly native plants,** which have co-evolved with many other plants and animals and together create a biodiversity that adds to the stability and re-silience of all (see Section IV for more about ecosystems). Select a variety of species so that there will be flowers and fruit throughout the growing season. In southern regions, this could mean creating a landscape that is flowering and fruiting on a year-round basis.

Actions beyond our own boundaries

There are plenty of ways to help wildlife move around in the world beyond the boundaries of our own properties. These actions generally take form of sup-porting or volunteering with local or region-wide organizations that work to protect or create natural areas and expand habitat connectivity. Such projects might include:

- Urban greenways and greenbelts
- Adaptive reuse projects such as the High Line Park in New York City

Many species are already changing their movements due to climate change, right through our human habitations. Although major migration corridors will still require protected open space, it's a good idea to design yards so they contain multiple habitats, whether big or small, to create various opportunities for species to live and move through. For the small landowner, it's not necessary to plan out whole connected corridor systems, but rather providing havens of native species will do a lot.

— Dr. Julie Richburg, Ecologist, The Trustees,
a Massachusetts land trust organization.

What about Assisted Migration?

Many plants and animals have shifted their ranges northward in recent years, on average about 4–10 miles (6–16 km) per decade. Spring events such as plant blooming, frog breeding, and bird migrations are happening several days sooner than they used to. With so many plants and animals under stress, it's natural to wonder: should we help out in our gardens and yards by planting some species in locations further north than they would naturally occur?

At this time, most conservation biologists say *no*. They caution that the risk of damaging existing ecosystems is too great, that the chances of accidentally introducing an invasive species, and doing more harm than good, are too high. Plus, we don't really know what will happen—how species will respond to the changing conditions.

For example, ecologists have predicted that with warming temperatures, cold-adapted species will move either northward or upward to higher elevations. But new studies reveal that many tree populations in the US, in particular deciduous trees, are tending to move *westward*. (Of course, the trees themselves don't move. But when their progeny [saplings] consistently turn up in greater numbers in new locations, this is considered movement of the population as a whole.) Some theorize that these shifts are driven by moisture considerations, not temperature. But we don't know for sure.

The act of intentionally moving species to a new location is a serious step that should not be undertaken casually. Therefore, in this book we strongly recommend that the general public not engage in relocating species northward with the express goal of helping them survive. Instead, we encourage gardeners to continue to purchase plants—ideally native plants—that have been identified as belonging in their own local Plant Hardiness Zone. These zones will be adjusted, as needed, as conditions change (see II-2 on page 58 for more about Hardiness Zones).

In addition, we encourage everyone to increase the extent of semi-wild (or even totally wild!) spaces within yards, neighborhoods, towns, and cities everywhere.

- Natural areas in parks, set aside from athletic fields and other recreation areas
- Native plant gardens in urban arboreta
- Regeneration/restoration of riverbanks and floodplains
- Wildlife bridges over or under highways
- Local nature centers
- Cluster development and other planning policies that consolidate development

Last Thoughts

Habitat corridors are key to protecting landscapes, helping plants and animals move around, and providing spaces that invite people into nature. All of these elements are especially important in urban areas, where uncountable acres of buildings, roads, parking lots, and highways block movement between the limited natural areas that do exist, both within and outside the municipal sprawl. And during this era when many species may need to relocate to more suitable habitat, cities often present unlivable, uncrossable, and even fatal barriers to this essential movement.

> Based on the most recent research, we really have no idea how wildlife will react to climate change. At the forefront of the battle every simple action makes a difference. Every yard, every park, and every open space would do well to be planted with native species that support wildlife by establishing forage, shelter, migration, overwintering, and escape habitats.
>
> — Bill Lattrell, Restoration Ecologist and Adjunct Professor at The Conway School

Credit: SReed

Figure VIII-15: Every suburban landscape can be designed to contain a rich abundance of niches and habitat to help support wildlife on the move.

IX FOOD

Overview

Growing food in home and community gardens is gaining in popularity for many good reasons. Small-scale farming and edible gardening appeal to those who want to:

- Increase local food security
- Grow food to feed their families
- Support local economies
- Cultivate heirloom and regional crops
- Reduce the encroachment of agribusiness on forests and other vital ecosystems
- Reduce their own carbon footprint
- Save money

Note: This section is not a tutorial on growing food crops. Instead, here we present broad ideas, applicable in all but the most arid regions, to help plan for growing more food locally.

> What we eat is responsible for a whopping one-third of all atmospheric warming today. Global meat and dairy production together accounts for roughly 15 percent of total greenhouse gas emissions, making the livestock industry worse for the climate than all the world's planes, trains, and cars combined.
>
> —Anna Lappé, "Eating on the Brink: How Food Could Prevent a Climate Disaster," May, 24, 2017, civileats.com

Urban Food Production

Every pound of food grown in home or community gardens reduces greenhouse gas emissions by two pounds.

"Urban food production could reduce net greenhouse gas emissions compared with the conventional food system because of its potential to produce food with lower intensity of transportation, energy use, and packaging, and greater carbon sequestration. More indirect ways in which gardening might reduce emissions include reducing or replacing urban lawn, and reducing energy consumption for air-conditioning, computers or driving due to more time spent outside."[1]

Climate Change and Food

Numerous studies have analyzed the ecological footprint of our typical food supply chain: from commercial farms to grocery stores and onto our dinner tables. Although the problems are substantial on many levels, as individuals we can make many climate-wise choices. In addition to growing at least some of our food and composting our kitchen scraps, we can eat less meat and buy more of our food from local sources such as farmers markets or CSAs (Community Supported Agriculture). We can frequent restaurants that feature locally grown food on their menus, donate leftover food to organizations that help the needy, and donate waste for composting and use at local farm operations.

In the book, *Drawdown: The Most Comprehensive Plan Ever Proposed to Reverse Global Warming*, editor Paul Hawken ranks reduction in food waste as the third most effective strategy out of 100 listed for reducing CO_2 emissions: "If 50 percent of food waste is reduced by 2050, avoided emissions could be equal to 26.2 gigatons of carbon dioxide. Reducing waste also avoids the deforestation for additional farmland, preventing 44.4 gigatons of additional emissions."

In This Section

The following *Action Topics* present ideas for growing food in a time of climate challenges:

- IX-1: Grow Food Above the Ground
- IX-2: Grow Long-Lived (Perennial) Sources of Food
- IX-3: Use Organic Methods for Growing Food
- IX-4: Support Locally Produced Food
- IX-5: Harvest Wild Edibles

ACTION TOPIC **IX-1** **Grow Food Above the Ground** »

Why This Matters

From the perspective of climate change, growing plants above the ground—in raised beds or containers—offers several advantages over the conventional method of growing directly in the ground. The most important of these is that when the earth is not turned or disrupted with plow or tiller, the integrity of the soil ecosystem is preserved. This maximizes its carbon-storing capacity. Additional benefits include:

- Less water is needed because water only falls on the growing areas with drip, micro-spray, or hand irrigation techniques.
- Non-growing areas are likely to have fewer weeds because they are not irrigated, which means that soil is not disrupted by pulling weeds, and therefore can continue to store carbon.
- Soil where crops are growing is never compressed by foot traffic because beds are sized for maintenance from outside the beds.
- Crops are planted intensely with just enough space to grow so weeding is less of a chore.

Figure IX-1: This raised bed is neatly maintained, and woodchip mulch on the walking areas protects the soil.

Credit: GStibolt

Actions

◗ **Pick the right spot.** A sunny, non-soggy location is best for an edible garden (except in warm climates where high shade may extend the growing season for some cool-weather crops). In general, crops that fruit, such as tomato, squash, and beans, need more sun than leafy or root crops. Be sure to consider roots of nearby trees and shrubs, which can spread far beyond their drip lines and will outcompete vegetables for water, even in a raised bed.

Another factor in locating beds or containers is access to an irrigation source, whether it be rain barrels or an outside spigot. (See III-2 on page 80 for irrigation ideas.) Keep in mind that food gardens can be in the *front* yard, as was common in days long past. But also keep in mind that some municipalities and/or neighborhood regulations may limit what's allowed in front yards. If that is the case, work to change those terribly outdated norms.

◗ **Use raised beds.** The sizing of the beds should allow people to work the whole bed from outside the bed; this usually means that beds are no more than four feet across—and less than that if the bed is accessible from only one side or if children will be doing some of the gardening.

Raised beds in contact with the ground do not need to have hard sides, and not constructing sides is the most climate-wise and most flexible way to go. A bed without hard sides can be any size and can be easily expanded, reconfigured, or modified. The soil mound will have sloping sides, which will need a stable mulch of some type to help hold it in place and keep the weeds down. Beds with no sides tend to drain better, which is better for most crops.

Hard-sided raised beds are usually built from inexpensive softwood like pine, which will last for several years, or from more rot-resistant cypress or redwood. (*Note*: Do not use pressure-treated lumber for an edible garden.) Composite lumber manufactured from sawdust and/or plastic may be the longest-lasting lumber, and, because it uses recycled waste product, may be a greener choice (see X-2 on page 274 for more about both pressure-treated wood and plastic lumber). Cinder blocks are somewhat alkaline and may slightly raise the soil pH, but many people use them quite successfully for their edibles. (See Section X for a discussion of materials used in landscape construction.)

Plan for drainage, because intense storms may produce several inches of rain in a day, which could wash away crops and soils where there is no planned drainage. In addition, saturated soil impairs root growth, which can kill some plants. To help prevent soil saturation when there are hard sides, plan for gaps or drainage holes near the bottom of the beds. For raised beds without hard sides, arrange the crops in wide rows with trenches between them for drainage. *Note*: When soil smells sour or like rotten eggs, it means that the soil has become anaerobic and potent greenhouse gas emissions (either methane or nitrous oxide) have developed.

Figure IX-2: Raised beds without hard sides can be any shape, and will drain better than those with hard sides. This bed has been mulched with pine needles raked from the neighborhood streets. Here, one crop had just been harvested and the soil had been enhanced with compost.

⏵ **Provide enriched soil.** Except for legumes (beans and peas), it's best to provide rich soil for crops in raised beds by adding composted manures to the general compost mixture. The most sustainable type of manure depends on what's available in your location. If you have access to fresh manure such as horse manure plus the bedding straw, compost this separately from your general compost pile for at least three months layered in with straw, leaves, wood chips, or other readily available organic material to help keep it from becoming anaerobic. (See V-4 on page 144 for more about compost.)

Building soil in place (see below) has a much lower carbon footprint than buying soil—especially bagged soil. However, buying soil may allow for a faster

Figure IX-3: This new bed has been started by layering about six inches of woodchips, then six inches of dried leaves, then four inches compost, and finally four inches of leaves. The layers will keep the bed moist, but well drained, and should sit for at least a few weeks before being planted.

start to growing vegetables, so it may be worth it, climate-wise. Purchase organic soil with no "moisture balls," no artificial fertilizer, and no peat moss. If possible, buy it in bulk rather than in bags.

In most situations, the underlying soil will not need to be tested because new soil will be built or added on top of the existing soil, but if the site is urban and/or if the history is not known, test the soil for pollutants such as heavy metals. If toxins are found in the soil, it's prudent to completely separate the soil in the raised beds from the underlying soil by raising the beds up on legs so there is space between the beds and the soil—or by using containers. Also, it's a good idea to cover the soil around the beds with a thick layer of woodchips or other persistent mulch so toxins are not scuffed up from foot traffic.

❯ **Create straw bale gardens.** You can create a raised bed garden by planting directly in one or more bales of hay or straw. This method may be ideal for crops that have had problems with soil-borne pests. Straw bales are usually preferable to hay bales because they contain fewer weed seeds, but either option is good for areas with low-quality soil.

Lay the bale flat, then with a garden fork dig out a hole in the middle of the bale that's about one half the surface. Fill the hole with your compost and soil mixture. The depth of the hole depends upon what you are planting. For a larger straw bale garden, arrange several bales and tie them together with twine. Before planting, let the bales sit for several days, and water thoroughly each day. After planting, keep a close eye on the moisture—you may find that these gardens require more frequent irrigation. At the end of the growing season, the straw and soil mix can be added to your compost pile or used as mulch.

❯ **Create container gardens.** Growing edibles in hanging or free-standing containers is the only way to produce a harvest on a balcony or patio, but container gardens can add quite a bit of variety and options to complement more traditional edible gardens. For instance, you may wish to contain mint and other aggressive spreaders in pots.

Using containers also provides the option of relocating the containers to shadier locations to extend the season for cool-weather crops; to sunnier locations to extend the season for warm-weather crops; or inside to prevent damage from frost for tender crops. This would be a temporary location, and the containers would be moved outside again, at least during the days.

Despite what we've been told all these years, a layer of gravel or potshards in the bottom of a container actually *impedes* drainage. Since plants are under enough stress in containers, increase the volume of the soil in the pot by not using gravel. Prevent soil from washing out of the drainage holes by placing a piece of screen or non-woven weed barrier cloth, or even a few dried leaves or pine needles over the holes before adding the soil. For the best drainage, use a tall pot.

Credit: GStibolt

Figure IX-4: Crops growing on this green roof include squashes, cucumbers, tomatoes, okra, eggplants, and peppers. This roof also contains a pollinator garden and an array of solar panels.

❯ **Grow crops on green roofs.** Many green roof systems are designed to use a minimum of soil (or no soil), to save on weight. This is an increasingly important option for growing food, as many urban neighborhood organizations are sprouting up to provide rooftop-grown food for local residents. See VIII-2 on page 226 for more on green roofs.

❯ **Consider growing crops in water.** Hydroponics and aquaponics are similar in that crops are grown with no soil, using just nutrient-rich water. These systems are mostly used for relatively small leafy crops such as lettuces, but many types of crops can be grown in this way. The space required to grow these crops is quite a bit smaller than soil-grown crops, and, since the water is re-circulated, the crops are

Credit: GStibolt

Figure IX-5: An urban hydroponics system supports both horizontal beds and stacked hanging baskets that form a green wall. Systems such as these use far less water than soil-based growing methods.

produced with about one-tenth the volume needed for soil-based gardening. In hydroponics, circulated water is enhanced with nutrient-rich additives. Aquaponic systems incorporate fish, and the resulting water, fertilized by fish waste, is used for growing plants.

Last Thoughts

There are many ways to grow food above the ground. Care must be taken to provide adequate irrigation since the volume of soil may be smaller with some of these methods. You can produce an amazing amount of food in just a small space.

ACTION TOPIC IX-2 Grow Long-Lived (Perennial) Sources of Food

Why This Matters

When crop plants produce for several years or longer, this means that:

- Once the plants are established, the underlying soil will remain undisturbed for years. The integrity of the soil's ecosystem is preserved and the soil continues storing its carbon.
- These plants, especially trees and shrubs, can cool the air and serve a larger role in local ecosystems.
- The sheer volume of food produced over the life of persistent crop plants provides more return for the investment, both financially and environmentally.
- Roots of perennials, shrubs, and trees will spread much more than annual or seasonal crops, which will help them to be more drought-tolerant.
- The rhizosphere (root zone) has a chance to become more established, so the soil's ecosystem can supply most of the needed nutrients. Reduced demand for fertilizer will shrink the climate footprint.

There are quite a few edible plants that are perennials: garlic, chives, oregano, mints, rhubarb, asparagus, and strawberries last several years or longer. There are also many shrubs that produce edible fruit such as blueberry, raspberry, blackberry, serviceberry, pawpaw, huckleberry, and hazelnut that will live for ten or more years. Trees that produce fruit and nuts—apples, pears, peaches, citrus, pecans, walnuts, and almonds—can produce for 20 years, or even much longer. (For information about wild edibles, see IX-5 on page 264.)

Actions

▶ **Plant herbaceous perennial crops.** Choosing a planting site for a long-lasting crop will depend to some degree on how the annual or seasonal crops are arranged. If raised beds are used, they may need to be somewhat deeper for crops such as asparagus or rhubarb. Keep in mind that these plants will spread and will not be part of the normal crop rotation. Soil disturbance around them should be avoided as much as possible—except

Credit: GStibolt

Figure IX-6: Blueberries are a favorite crop for regions where there is enough rain and somewhat acidic soil.

It's easy to think that all the energy for your garden comes from the sun, but that's not usually the case. Home gardeners seldom consider the fossil fuel energy it takes to grow a garden, but when you add up the energy inputs required to obtain materials, prepare the soil, plant the crops, water and fertilize them, and preserve the harvest, it's surprising how much outside energy your garden can end up consuming.

—Owen E. Dell, Landscape Architect, "Ten Tips for an Energy-Efficient Food Garden," 2017

Credit: GStibolt

Figure IX-7: This herb garden with perennials—chives, meadow garlic, and rosemary—provides food without causing soil disturbance.

Support Pollinators

Most fruit and nut species require pollinators—mostly bees—to fertilize their flowers and enable fruit formation. Commercial fruit orchards use honeybees (native to Europe), either by raising them or by hiring itinerant bees during the blooming cycle. For smaller operations, encouraging native bees works well. Support native bees by planting pollinator gardens and hedgerows to provide nectar and pollen when the fruit trees are not blooming. It also helps to leave some soil unplanted and un-mulched so the bees can dig their nests. (And, of course, avoid using pesticides!) See VII-2 on page 205 for more on pollinator-friendly habitat.

for annual or semi-annual compost topdressing for crops that benefit from rich soil. As an alternative, some people include these crops as part of perennial borders in the landscape. (See VII-4 on page 214 for perennial border ideas.)

Before planting most perennial crops in the ground, enrich the soil with a top-dressing of compost and composted manures—it's best for the soil if the amendments are not dug in, and you should apply them when no hard rain is forecast. Cover the area with persistent mulch, except right around the new crop seedlings or starts. Over the years, more enrichment may be necessary to maintain vigorous growth, depending upon the crop. (See IX-3 on page 257 for details on organic methods for soil enrichment.)

● **Plant shrubs and trees that produce edible fruit and nuts.**

Note: While this book advocates planting native species, there can be advantages to planting non-native trees and shrubs, especially food-bearing species. Be aware, though, that these may need more irrigation, more pruning, and other special care to ensure their crop yield.

Fruit and nut trees

Because fruit and nut trees are often quite expensive, choose only those that work well for the local climate and avoid those that are under threat of a known ongoing disease (such as greening disease for citrus crops in Florida). By giving yourself the best odds for success, these crops will live out their expected life spans and produce for many years. The local extension office should have up-to-date information on the best tree and shrub crops suitable for the local area.

Many fruit trees are grafted onto small-tree rootstock to check their growth, which may make them easier to harvest and maintain in the long run. (See Section II for more about trees and shrubs in the landscape.)

Some nut trees such as almonds require substantial irrigation, while others such as pecans are more tolerant of drought. Find out how much irrigation is recommended before you plant them and if the need is great, see if it's possible to irrigate with greywater. (See III-2 on page 80 for more on greywater systems.)

Fruit-bearing shrubs and brambles

Throughout North America, many species produce edible fruit for humans, and simultaneously provide many ecosystem services. Blueberry, huckleberry, pawpaw, elderberry, cloudberry, serviceberry, bunchberry, black-currant...the list goes on and on. When possible, use species native to your area or cultivars that are based on the native species. For example, New Jersey blueberries are fantastic and tremendously fruitful in their region, but they are unlikely to grow well in the South or West.

To maximize harvest, some people use fine netting to keep birds and other wildlife away. However, you shouldn't use it during pollination periods. Consider planting some of these important understory shrubs just to support birds and other wildlife.

In many regions, it's possible that edible fruit-bearing brambles such as blackberries or raspberries may sprout naturally in places where landscapes are allowed to "go natural." If this happens, take advantage of their presence and cultivate them to maximize the harvest. While brambles are technically

Figure IX-8: A patch of brambles like these blackberries needs very little care in the right environment.

Credit: GStibolt

Permaculture

The term *permaculture* originated with Australians Bill Mollison and David Holmgren in the 1970s to denote *permanent agriculture*. The concept has now also come to be called *forest gardening*. Using more trees, shrubs, and perennials is part of the philosophy, along with maintaining closed-loop cycles of all materials to keep the whole system of growing food, including animals, in balance. *Note of clarification*: Forest gardening does NOT mean growing food in existing, established forests. Rather, it signifies food grown in multiple vertical layers for maximum productivity, like a forest. In the permaculture approach, nothing is wasted, and instead all "waste" is re-imagined as a surplus resource, to be used as an input into another process.

Over the years, permaculture's many supporters have redefined the term to mean *permanent culture*, a concept that embraces many interlinked aspects of sustainable living that go beyond just growing food. Permaculturists promote many positive actions and have changed the way many people think about the usefulness of a landscape. This philosophy appeals to a growing number of homeowners and gardeners.

A caveat: Some permaculture practitioners give short shrift to the problem of invasive plants. Instead, they believe that any edible or useful plant is worthy of cultivation, regardless of its invasive potential.

considered shrubs, they do not produce wood. Instead, they grow in 2-year cycles, with the first year's growth, called a *primacane*, being only vegetative. In the second year that same cane, called a *floracane*, will bloom and bear fruit. These plants spread aggressively, so only grow them where they have plenty of room—away from your other edibles and high-traffic areas.

Last Thoughts

Many crops are perennials, shrubs, or trees, and adding them to the landscape makes it more useful, productive, and climate friendly. Because long-lived crops grow from established roots, they require less time and energy than seasonal/annual crops. When soil is less disturbed, microbes that help the plants be more efficient in absorbing water and nutrients are preserved, so plants are generally more drought tolerant.

ACTION TOPIC **IX-3** **Use Organic Methods for Growing Food** »

Why This Matters

Organic methods include taking care of the soil's ecosystem by adding more humus to the soil to increase nutrients for crop plants, and avoiding the use of poisons of all types—whether synthetic or organic, homemade, or commercially made.

Note: This book does not cover the organic certification process, which allows growers to use the label *organic* or *organically grown* when selling their products.

There are many reasons why organic gardening methods work well for growing crops—in larger community gardens as well as in a small family garden. Organic methods:

- Enhance and balance the soil's ecosystem, which helps the crops to thrive.
- Regularly add organic materials to the soil so it holds more moisture and stores more carbon.
- Include non-poisonous, physical, and manual pest control, along with working with natural predators, to reduce crop damage.
- Allow growers to take full control over their food from seed to table.
- Reduce greenhouse gas emissions by using local organic materials to enrich soil instead of manufactured fertilizers that are bagged and transported great distances.
- Work with natural systems, so crops become part of the local ecosystem.

But, mostly, people appreciate organically grown food because it involves working *with* Mother Nature, not against her—by building fertile soil naturally, inviting beneficial bugs and other organisms to the garden, avoiding the use of toxic chemicals, recycling plant materials, and providing ideal conditions for particular crops. In addition, many people think that organically grown foods are tastier than conventionally grown food.

Actions

- **Become an educated grower.** Learn which crops are most likely to succeed in your area and the best times to plant them. The local agricultural extension office should have data sheets for local crops. Extension offices often provide workshops or other advice to help beginners and people new to an area to make the best choices. The advice found in general vegetable gardening books may not work well for all locations: getting local information is key to success.

- **At first, choose the easiest-to-grow crops.** When just beginning to grow food, or growing food in a new location, choose plants that will thrive in your climate with a minimum of care and without the use of poisons. These will probably include those that have been bred to resist diseases such as fungal wilt and other common problems.

  After gaining some local knowledge and experience, add a few new or unusual crops to the mix each season. Many people tout heirloom crops as the ones with the best taste, but it's possible they are vulnerable to a pest or disease—which may be why the mainstream seed companies stopped offering them; so keep an eye out to catch any problems early. Of course, the best crops are those that you and your family will enjoy eating.

Figure IX-9: Cool-weather carrots are an easy-to-grow crop that most people, even kids, like to eat. This bunch includes some red carrots in addition to the traditional orange varieties.

Credit: GStibolt

▶ **Find local sources for manures and other organic matter.** Compost made of local materials works well for non-edible gardens and the wilder sections of the landscape, but crops generally require richer soils because we're pushing them for fast, bountiful harvests. Seek out local sources for manures from herbivores (horses, goats, chickens, cows, or rabbits), but not dogs, cats, or humans. Sometimes local mushroom growers have mushroom compost available. Manures that have been packaged and shipped to stores will work well in edible gardens, but their large carbon footprint makes them a less climate-friendly choice.

> ## Consider Growing Fungi
> Fungi are not plants: they are *saprophytes* that get energy from digesting animal and plant matter. The mushroom is actually a fruiting body that produces spores to start the next generation. Edible mushrooms are easy to grow, but it's probably best to start with purchased spores for the type of mushroom you like and then inoculate a *substrate*, which will need to be kept damp while the fungi grow. The growing medium could be a log, wood chips, composted manures, or even coffee grounds. There are many online resources and excellent books on growing fungi as a crop.

▶ **Rotate crops and use cover crops.** Growing the same seasonal crop in one space year after year will cause nutritional deficiencies in the soil and a concentration of soil-borne pests of that particular crop. Crops in the same plant family should be considered as the same crop. For instance, planting summer squash followed by cucumber or watermelon would not be an effective crop rotation (they are all in the squash family). A better plan would be to follow that summer squash with a bean or pea crop, and then a lettuce. Learn which crops are in the same plant family and then treat them as a group.

It's also a good practice to grow a cover crop of some type as part of the mix. Cover crops are generally not grown to be harvested, but are dug into the soil a few weeks before the next crop is planted. Cover crops are often members of the legume family, which enrich the soil, but other choices include winter grasses, marigolds, or sun hemp.

▶ **Plant pollinator-attracting plants among or near the crops.** While some crops are self- or wind-pollinated, many rely on pollinators, primarily insects. For instance, members of the squash family bear separate male and female flowers and require 7 to 10 visits from pollinators for a fruit to form. Building a pollinator garden near the crop plants will increase the yield. And don't forget about

Credit: GStibolt

Figure IX-10: Squash blossoms attract a variety of pollinators with their nectar. The pollinators then carry pollen from male flowers to female flowers. Here, three bees, all different species, feed on (and pollinate) this huge Seminole pumpkin flower.

the fruit trees, because most of them need pollinators, too. (See VII-2 on page 205 for details on attracting pollinators.)

❯ **Minimize the use of poisons.** Sometimes the standard instructions for home and community gardeners, especially if they are growing fruit trees, includes the preventive use of dormant oils and other poisons to combat possible pest infestations. Many of the suggested poisons are derived from plants such as neem trees and chrysanthemums, but a poison is still a poison. Using organic methods means that poisons, even organic ones, are used rarely and *only* as a last resort to reduce crop damage. Usually, manual and integrated pest management methods work well for home or community gardens. (See "A Primer on Landscape Chemicals" in Section I for a description of the poison cycle and how to step out of it.)

Last Thoughts

Using organic methods for growing food helps to mitigate climate change in several ways, including taking care of the soil. In addition people appreciate knowing that food they grow themselves is safe and nutritious to eat.

> Odd as I am sure it will appear to some, I can think of no better form of personal involvement in the cure of the environment than that of gardening. A person who is growing a garden, if he is growing it organically, is improving a piece of the world. He is producing something to eat, which makes him somewhat independent of the grocery business, but he is also enlarging, for himself, the meaning of food and the pleasure of eating.
> —Wendell Berry, in *The Humane Vision of Wendell Berry*, 2008

ACTION TOPIC IX-4 Support Locally Produced Food »

Why This Matters

Buying locally grown and produced food helps support local farmers and economies, and it reduces food's ecological footprint due to transportation from distant locations. In addition:

- Food that is transported great distances is usually picked before ripening and allowed to ripen in-transit, while produce that is allowed to ripen in place is likely to have more nutrients.
- Growers who sell food locally may choose varieties based on taste rather than on shelf life or ability to withstand rough handling.
- Locally grown food is handled by fewer people from seed to table than commercial food, which reduces damage and waste. (See sidebar on food waste.)
- Food that is harvested and sold locally within 24 hours of harvest is unlikely to be packaged, which reduces its climate footprint.

Credit: GStibolt

Figure IX-11: Squashes with flowers still attached were picked that morning. This type of fresh offering is usually only available at local farmers markets.

Actions

In addition to growing edibles on your own property or in a personal plot in a community garden, there are a number of ways to be a *locavore*—one who eats locally grown food.

⟩ **Participate in fleet farming.** Fleet farmers use other people's yards to grow food. Homeowners/landowners who don't have the time or know-how to grow vegetables sign agreements with farmers—who will do all the work. Often, the farmers work in one neighborhood and ride bikes between the yards with all the tools carried in bike trailers. Financial arrangements vary, but usually the homeowners pay an initial fee, provide water for irrigation, and receive some

Food Waste

Food waste is not only an economic problem, it's a huge environmental problem—in part, because so much perfectly edible food ends up in landfills where it emits methane as it breaks down. In the book, *Drawdown*, editor Paul Hawken ranks reducing food waste as the third most important strategy in reversing climate. In addition, many hungry people could be fed if we diverted perfectly edible, but imperfect, food from the waste stream.

Food waste occurs in the fields where misshapen food is left unpicked, in stores, in restaurants, in home and community gardens (when too much food is produced), and in home kitchens. There are ways we can reduce this waste.

- *After-harvest harvests* occur when community groups or other organizations are allowed into fields after the farmers' workers are finished harvesting the "saleable" produce. They harvest the less-than-perfect foods and distribute this second harvest to various charities or other groups that find ways to use this perfectly edible food.
- Composting groups collect food waste from restaurants, compost it, then sell finished compost to local farmers and/or donate it to community gardens.
- For people who have grown too much for their own use or who have fruit trees that produce much more than they can use or are able to pick, there are groups who will come in and harvest the fruit and distribute it to charitable groups or sell it at farmers markets. Various online organizations match up local pantries with people who have more than they can use.
- Our personal food waste, from both preparation and leftovers, can be added to a compost pile, or it can be incorporated directly into a vegetable garden by using trench composting.

Figure IX-12: Farmers markets offer mostly locally grown produce, often including unusual varieties that are never seen in grocery stores. Customers also have the chance to talk to the growers so that they have a better connection to their food. In addition, locally grown food has a very small carbon footprint.

Credit: GStibolt

of the harvest. The farmers then sell the rest of the produce at local farmers markets. Some also donate a portion of the harvest to charity.

A similar arrangement is increasingly happening in many cities, where apartment-dwellers engage an individual or farming group to grow crops on a green roof and deliver the harvest to residents of that building or neighborhood.

▶ **Buy into a CSA (Community Supported Agriculture).** Local farms pre-sell shares of their harvest. Individuals pay a fee at the beginning of the season and receive a box or basket of food once a week during the growing season. If a particular farm includes some unusual vegetables as part of the harvest, they may provide recipes as well. This arrangement shares the risk of farming across the community, and the farmers have money to pay for seeds, labor, and supplies before the season begins.

▶ **Buy from local farm stands and farmers markets.** Small family farms can better survive if they sell some of their foods directly to the public at their own roadside farm stands and/or by participating in local farmers markets.

▶ **Buy into an agri-hood.** Some communities are built around an organic farm with one or more resident farmers, instead of a golf course, say. The organization of these communities varies but often includes the option of attending educational workshops on growing and/or cooking the food, working on the farm, buying into a CSA, shopping at farmers markets of the community farm, or other local farms. In some cases, a farm-to-table restaurant is part of the mix. (See below.)

▶ **Patronize local farm-to-table restaurants.** Restaurants and caterers that are associated with one or more local farms provide locally grown food items on their menus. Some suburban or rural restaurants have gardens right on their properties that are supplemented by local farmers. Online directories can help with finding these restaurants. Some schools have also been set up with farm-to-table food for their cafeterias.

> The biggest thing you can do is understand that every time you're going to the grocery store, you're voting with your dollars. Support your farmers' market. Support local food. Really learn to cook.
>
> —Alice Waters, in a 2014 interview with Amy Cavanaugh

❯ **Buy from supermarkets that grow food on their roofs.** The average size of supermarkets is a little less than an acre, which means that there is a lot of space on roofs to grow food. Some stores have hired companies to install greenhouses and grow food right on their roofs, often using hydroponic systems. The arrangements vary, but this cuts out all the transportation costs, and the greenhouse operations can be run using only solar panels.

ACTION TOPIC (**IX-5**) **Harvest Wild Edibles** ❯❯

Why This Matters

The world is full of wild plants that produce fruit, flowers, seeds, leaves, roots, and stems that humans can safely eat. Gathering food from wild sources makes sense and is climate-wise for several reasons. It can shrink the carbon costs of cultivating and irrigating gardens, and it will reduce the transportation footprint associated with buying store-bought food. Also, when unwanted weeds that you need to pull anyway are also edible plants, you can combine two gardening actions into one. And finally, harvesting wild foods saves time and energy compared to growing plants. It lets you skip all the other phases of growing food and just go to the last step: the harvest!

Figure IX-13: Meadow garlic (*Allium canadense*), often pulled as a weed, can be used like chives or garlic. This wild edible grows naturally throughout the eastern half of North America.

Credit: GStibolt

Note: Here we do not recommend specific edible plants, nor do we offer advice about how to harvest, store, or prepare food collected from wild plants. Many other books and websites can provide this detailed and regionally specific information. Rather, our aim is to bring this subject to your attention and encourage you to add wild edibles to your life, if possible.

Actions

❯ **Eat (some) weeds.** Weeds are plants that sprout up in places where they weren't planted and aren't wanted. The most climate-wise approach to weeds is to limit weeding activities to your edible beds and other areas where the weedy plants are crowding out more desirable plants. But many plants that are removed from gardens and landscapes are actually edible. (*Note*: This is true long as they haven't been sprayed with pesticides. See "A Primer on Landscape Chemicals" in Section I for more information.)

Figure IX-14: Dollarweed or pennywort (*Hydrocotyle umbellata*) is an aggressive weed especially in damp areas. It can be used in salads, stir fries, pestos, and dips.

Before gathering wild plants to eat, it is essential that you positively identify the plants and know exactly what you are harvesting. Many plants, even those in the same family as edibles, are toxic— some extremely so. To avoid any confusion, don't depend on common names; make sure you also know the scientific names. Local extension offices can help with plant identification. Take a stem with several leaves, a photo of the whole plant, and if possible, a flower or fruit, to be examined.

If you're pulling edible weeds from your garden beds or other parts of your landscape, treat them as you would any other harvested plant. Pre-rinse in rain-barrel water and keep them cool until you're ready prepare them.

If you are aware of a bunch of sprouting seedlings that would become a mass of weeds if allowed to grow, wait until you're ready, and then harvest them as microgreens. You'll learn what seedlings look like by paying attention to volunteer plants over the seasons.

❯ **Encourage edible weeds.** While you may wish to remove most of the populations of weedy-looking plants to keep your landscape looking neat, you could also intentionally make room for some of the edibles, even in little pockets or corners of smaller landscapes, so that you will have a ready access to this free food.

You might also want to cultivate wild plants as part of your regular gardens. For instance, you could plant some meadow garlic next to your chives or garlic chives, which are perennials used for similar purposes. Also, as stated in Action Topic IX-2 in this section, you may wish to plant a patch of brambles so you have your own stands of blackberries and raspberries.

❯ **Forage for wild edibles.** If you have the opportunity to harvest food from natural areas, first be sure that you have made a positive identification of the plant. Remember that there are many look-alikes in nature, so be very careful. Then before you start, observe these important guidelines:

* Get permission of the landowner, even if it's a right-of-way or nearby community property. Also, while you're communicating with the landowner or manager, make sure that the area has not been sprayed with any toxic chemicals.
* Wherever you are collecting, take no more than 10 percent of any population or stand. Also, gather from the middle of a population, not from the edges, where the plant's roots are actively extending the size of the patch.
* Collect only what you can use, and don't collect wild edibles for wholesale-to-retail markets. Wild populations (of both plants and animals) are already stressed by climate change and loss of habitat. We should not unintentionally magnify that stress by damaging or destroying patches of ecosystems, even if they seem like "only weeds" to us.
* Never collect rare or endangered species.

Figure IX-15: Spotted beebalm (*Monarda punctata*) dresses up this meadow area with its flowers and many pollinators. In the mint family, this plant contains the same oil as thyme and oregano.

Credit: GStibolt

Check plant lists provided by local native plant societies, arboreta, or extension offices, to be sure.

- Avoid areas where you don't know the history of the land, especially in urban areas, because there may be toxins in the soil that have been absorbed by plants.
- Avoid roadsides because plants there are exposed to toxic vehicle exhaust and possible roadside spraying.
- Be especially careful when foraging for mushrooms. Some can be difficult to positively identify, and quite harmful if misidentified.
- Check local resources for lists and photos of known poisonous plants.

In many parts of the continent, native trees and shrubs provide nuts that humans, as well as the local fauna, find delicious and nutritious. Some of these are edible in their natural form, but others (such as acorns and beechnuts) require some kind of treatment to make them palatable to humans. In any case, this is free food falling from the trees. If you don't collect this bounty from nut trees that are growing in your own landscape, some nuts will sprout into new seedling trees, perhaps in places you don't want such trees. So collecting the nuts might serve two purposes—free food and pre-weeding. Remember though: it's important to leave a significant portion of what's available for the local wildlife that depends on that food.

Despite these many caveats, harvesting wild edible plants can be great fun and very rewarding. Even beyond the pleasure of saving money and helping the climate, foraging for natural food is an excellent way to connect to nature, enhance our diets, and broaden our tastes!

Last Thoughts

Throughout North America, nature provides a vast selection of plants that we could use to supplement our diets. These plants are well adapted to the climate in which they grow naturally, and need little or no assistance from us to produce abundant food (as well as medicinal substances). Why not take advantage of Mother Nature's bounty?

> Wild food plants grow in the presence and protection of each other. They flourish in soil that through the centuries has seen an accumulation of humus and plant nutrients. Plants growing in woodlands and hedgerows have no need for man-made fertilizers. Nature provides their every need.
>
> —Roy Genders,
> *Edible Wild Plants*,
> 1988

MATERIALS

X

Overview

We can purposely design and build our landscapes to incorporate climate-wise materials, which are characterized by:

- Durability and potential for long lifetime
- A small carbon footprint (see "A Primer on Climate Change" in the Introduction for more on this)
- Strong potential to be recycled/reused/renewed
- Local availability
- Minimal impact on wildlife habitat and corridors
- High capacity for carbon sequestration

Important note: Unlike the rest of the book, this Section does not provide *Action Topics* or *Actions*. Because every landscape construction situation is unique, and each one involves its own particular mix of options, costs, traditions, local materials, and regional expectations, specific action recommendations would be impractical. Instead, this Section presents information to help you understand and evaluate your options, from the perspective of shrinking your carbon footprint and doing well for the planet. We do not recommend one material over another.

Also note: The guidelines presented here focus on *design* choices and do not cover specific *construction* techniques.

General advice about choosing materials

Obtaining firm data for every product or material under consideration can be difficult—or impossible, in some cases. Yet even in the absence of exact figures, we can easily reduce our climate impact by following these suggestions:

- Choose recycled or *repurposed* materials.
- Give priority to locally made materials to support local economies and reduce transportation distances.
- Be on the lookout for usable on-site materials, either on your own property or from nearby construction projects.

All of these choices will help reduce manufacturing costs and emissions, minimize waste going to landfills, reduce strain on ecosystems (potentially including old-growth forests), and generally be good for the planet.

Life-Cycle Assessment (LCA)

Life-cycle assessments determine the environmental impact of a process or product by measuring the energy consumed during:

- Raw material extraction and procurement
- Production and manufacture
- Transport and distribution
- Use, repair, and maintenance
- End-of-life recycling or disposal

LCA information can be quite difficult to obtain. It requires examining data from diverse, scattered worldwide sources—about processes that can be complex and sometimes not well documented at the source. Inevitable uncertainties, disagreements, and gaps must be resolved. And then the data must be translated into uniform units of energy consumption (usually in joules) and CO_2 and pollution emissions (usually in gigatons or kilograms).

As a result of this complexity, data simply don't exist for most landscaping processes and products. One fact is true even without the support of studies and data: today in our Western society, we tend to use more of the materials with large carbon footprints and environmental impact (concrete, steel, and asphalt) than those with smaller impact (wood, earthen, and do-it-yourself products). It would be ideal if we could try to rebalance or even reverse that equation, either by using less of the materials with a large climate footprint, or by encouraging manufacturers to shrink the footprint of their products, and buying from those that do.

In This Section

The following topics explore the various attributes of the five types of material that are most commonly used in landscape construction projects, including:

- X-1: Stone
- X-2: Wood
- X-3: Metal
- X-4: Concrete
- X-5: Earthen Materials

Overview

When used as a building material, stone can exist in the form of flattish slabs split off from bedrock, irregular chunks blasted loose from bedrock, quarried stone that is cut and shaped into specific sizes, large fieldstone boulders that occur naturally in some regions (often as a result of glacial deposition), or as granular stone mined and crushed into various dimensions. Stone may be used to build aboveground structures such as retaining walls, freestanding walls, steps, and benches, or to create on-ground surfaces for walking, gathering, and even driving. It possesses the following climate-related attributes.

Durability

Stone, the substance itself, is among the most durable of all the materials we can use for landscape construction because stone generally doesn't rot, erode, or break down over time (or at least the process is so slow as to not matter). Stone structures and surfaces can also be similarly long lasting, if they are constructed properly, i.e., with the necessary site preparation, suitable base, and correct techniques. And they can—again, if properly installed—be completely maintenance-free for their lifetime.

On the other hand, stone construction that has been built poorly can be among the most expensive and energy-intensive to fix. This is because the job often requires the return of large, fuel-consuming and CO_2-emitting

Credit: SReed

Figure X-1: Cut granite steps are a familiar landscape feature, but be sure to build steps with proper dimensions for comfortable use.

Credit: SReed

Figure X-2: It's perfectly acceptable to use several types of stone in a single landscape. Here, driveway cobbles, cut granite steps, and modular bluestone pavers combine with local western Massachusetts Goshen stone to create an inviting arrival courtyard.

equipment, which can partly negate the main climate benefit of using this durable material in the first place. In the big picture, though, natural stone's extreme long life, along with its nearly universal aesthetic appeal, tend to outweigh these drawbacks.

Carbon Footprint

In an ideal low-carbon scenario, stone that is useful for landscape projects would be easily procured from locations within a few miles of your project. In reality, it is unlikely you will be able to do this. Typically, rock must be extracted from the ground using energy-intensive blasting or heavy excavation, or both. And since few rock quarries are conveniently located near residences (and sometimes not even for hundreds of miles), the energy cost and CO_2 emissions associated with transport can be very high. Sometimes this large *procurement footprint* can negate the climate benefits of using stone.

In addition, building with rock often requires the use of heavy machinery, both to prepare the site and to position the heavy materials. The site preparations might be needed in any case, but lighter-weight materials do exist that could serve instead of stone, to reduce the need for large equipment.

Stonework generally requires a level of skill (and strength) that few of us possess. So using stone for landscape construction might entail hiring a stonemason, thereby adding to the cost (not just our own out-of-pocket costs, but the mason's travel impact as well). Getting professional help, however, may be best way to minimize energy consumption and GHG emissions, and doing it right the first time can insure against the need for work to be repaired or replaced in the future.

> ### Do It Yourself?
> Many handy homeowners might wish to cut costs by trying to build with rock themselves. This can work in some cases, especially with paving rather than structures. For safety, though, we strongly recommend hiring a mason or landscape contractor for all stone walls or stairways over two feet tall.

Potential To Be Recycled/Reused

Natural stone is one of the most recyclable of construction materials. This is especially true for dry-laid stone; mortared stone can sometimes be too labor-intensive to disassemble and prepare for reuse (unless that reuse is to crush it into usable aggregate or dust). Salvaged, reused, or aged rock is often more desirable, from both environmental and aesthetic points of view, than freshly quarried, cut, or excavated rock.

Local Availability

Availability depends entirely on the location of your project and the nature of the stone you plan to use. In some parts of the US, native stone suited for landscape construction is extremely abundant, whereas other regions lack this resource entirely. In some places, wonderful rock can even be found lying on the ground or tumbling off old walls. *Note*: Some towns prohibit removing stone from historic walls and foundations.

Impact on Wildlife Habitat and Corridors

As with any kind of mining, the process of extracting landscape stone from the earth generally obliterates aboveground vegetation and the complex soil ecosystem that support it. This disruption is often made worse by waste materials stored in piles across the landscape.

Figure X-3: Redrock from a local quarry helps this Taos wall fit harmoniously within its rural setting.

Credit: SReed

Ideally, we should try to support quarrying operations that make an effort to minimize this harm and/or restore disrupted areas. In reality, this option is a heavy lift. The economics of stonework excavation and construction work against spending anything beyond what is absolutely essential for the procurement of the rock.

Carbon Sequestration Potential

The earth's largest carbon reservoir is in rocks, which hold about 66 billion gigatons of carbon. The vast majority of that is as a chemical component in regular rock like granite, sandstone, and limestone; a tiny bit (.004% of the total) occurs in coal, oil, or natural gas. (See "A Primer on Climate Change" in the Introduction for more about carbon.) However, despite this important role in the big carbon picture, rock doesn't actively sequester carbon on a timescale that's relevant in this discussion.

> Natural stone paving is an obvious choice for landscape design. Although it is not cheap, it is a beautiful, classic, and rugged material—and if sourced locally it can be a very sustainable choice because it requires few or no manufacturing processes.
>
> —Alice Bowe, *High-Impact, Low-Carbon Gardening*, 2011

Last Thoughts

Overall, using stone in landscape projects can be an excellent way to avoid using other materials with higher energy or transport costs. But the calculus, which may include factors such as availability/cost of a skilled installer, and of course aesthetic preferences, is unique to every locale, project, and individual. We strongly recommend using locally available stone in as many creative ways as possible.

MATERIAL (X-2) Wood »

Overview

Wood used for exterior construction is lightweight, yet strong. It is easy to work with using everyday tools and modest skills. And it seems almost infinitely adaptable to a wide variety of projects. For these reasons, in many landscape projects, wood can be the best choice of building material.

Durability

Wood can be a long-lasting or a relatively short-lived material. To maximize the lifetime of landscape structures built with wood, choose hard and/or rot-resistant wood; use proper construction methods; minimize moisture-absorption by applying a finish or preservative; and perform regular maintenance, both structural and surface.

Carbon Footprint

Wood is generally considered to be a carbon-negative product, i.e., it stores more carbon than it is responsible for emitting, and using wood avoids greenhouse gas emissions from other products. Building with wood also has a relatively small carbon footprint compared to other materials. Wood preservatives do have a large carbon footprint, so it's important to choose eco-friendly preservatives that are either organic or free of volatile organic compounds (VOCs).

Credit: SReed

Figure X-4: Using wood in long-term landscape projects is a great way to sequester carbon.

What about Pressure-Treated Wood?

For situations in which wood must be in contact with the ground, *pressure-treated lumber* (P-T lumber) is a widely used option. P-T lumber has been infused with chemicals that prevent insect damage within the wood fibers. The treatment also hardens the surface, making it more durable than most conventional lumber, and this durability can reduce demand on forests for replacement lumber.

On the other hand, P-T lumber is loaded with toxins. Do not use it in garden beds where food will be grown, or for deck surfaces where children might play (or adults walk in bare feet). Many landfills will not accept P-T waste due to the leaching of chemicals into the soil. Also keep in mind the added carbon footprint of manufacturing the chemicals used in the treatment, and the health risks of workers exposed to these chemicals.

Credit: SReed

Figure X-5: This quirky tree house built on a high stump demonstrates one way to make the most of wood for landscape construction.

Potential To Be Recycled/Reused

As a building material, wood has a high capacity to be reused for other projects. It is fairly easily disassembled with minimal harm to the original wood (and this process can be aided by purposeful construction with disassembly in mind). Salvaged wood is also relatively easy to find in some regions. Some people might recommend converting old, rotted, or decaying wood into mulch, but avoid using wood that has been painted or coated with any toxic preservative. And never use pressure-treated wood for mulch. Finally, old or unusable wood from construction projects could be used in local bio-fuel operations that also produce renewable electricity, thermal energy, and/or transportation fuels, ideally in combined heat and power systems.

Local Availability

Since wood suited for construction grows in so many different regions, one would think it would be easy to use locally harvested and processed wood products. However, the economic reality is that most wood travels hundreds of miles to be processed, and then hundreds more when distributed to the thousands of outlets where we can purchase it. If possible, try to support local lumber-processing operations and suppliers.

Impact on Wildlife Habitat and Corridors

Whenever possible, choose wood certified by the Forest Stewardship Council (FSC) or the Sustainable Forestry Initiative (SFI). These stamps indicate that a wood product comes from a forest that is sustainably managed, i.e., managed to restrict clear-cutting, protect fragile ecosystems, respect native cultures and economies, and prevent illegal logging. Both certifications may slightly raise the cost of the wood, but steady consumer demand will increase the supply, potentially bringing down the price.

Coppiced Wood: The Ultimate Climate-Wise Resource

Coppicing, a traditional European method of managing woodlands, involves cutting selected species of trees to their base, and allowing them to regenerate from the stump. The resulting young saplings can then be harvested to produce stakes, poles, and whips useful in making fences, raised beds, plant frames, and arbors. In this repeating cycle of re-growth and harvest, carbon is continually sequestered. Further, the coppice saplings provide excellent wildlife habitat and a source of insect food for birds and small mammals, as well as winter browse for large mammals. Taken together, all of these qualities give coppiced wood a very small climate footprint. *Note*: A similar process, called *pollarding*, involves pruning branches to a higher point on the trunk, not to the ground.

Credit: S eed

Figure X-6: Using local materials can be both climate wise and beautifull

What about Bamboo?

Bamboo's strength and flexibility make it an excellent building material for everything from houses to furniture to fences and more. And in recent decades, this fast-growing plant has also been promoted for its ability to take in large amounts of carbon from the atmosphere via photosynthesis and store it for up to 80 years. More research is needed to assess bamboo's potential as a carbon sink. But when it comes to using it in North American landscapes, any benefits gained from bamboo's carbon storage capacity are mostly offset by the large amount of energy required to transport it here from its distant sources. In terms of cultivating it for harvest, bamboo's potential invasiveness and disruption of ecosystems, many of them already made vulnerable by the stresses of climate change, are also of concern.

Credit: SReec

Figure X-7: The large transportation "footprint" of tropical woods, such as this green-roof decking and bench material, might be offset by their extreme durability, water-resistance, and freedom from preservative treatments, making their use climate-wise.

Carbon Sequestration Potential

Of all the building materials available to us, wood contains the highest potential for storing carbon. This is because the substance itself consists almost entirely of carbon. So when we preserve the wood intact, by using it instead of burning it or letting it decay, its stored carbon remains safely tucked away, out of the atmosphere. In fact, of all the building materials available to us, the only one that sequesters carbon is wood.

Note: Although wood sequesters carbon, we do *not* necessarily advocate cutting down trees to make decks and picnic tables. However, if wood is the material chosen for construction, its ability to be a long-term storehouse for carbon adds greatly to its benefits.

Last Thoughts

All things considered, wood ranks very high as a climate-wise material. Its main drawback is vulnerability to rot. Conventional preservatives that are used to extend the lifetime of wood may be harmful to both the environment and the people who work with the wood. We recommend minimizing the use of pressure-treated wood and instead using naturally rot-resistant wood, and designing projects to minimize the number and extent of places where wood will be in contact with soil.

> If demand for wood is too great and too impatient, the rate of harvesting outstrips re-growth. Quick harvesting limits the size of wood a forest can produce: where old growth forests once yielded huge beams, forestry today hurries to harvest 2 × 12s. With hasty harvesting, or where destructive methods are used, forest health declines. Push a renewable resource too far, and it faces at least local extinction.
>
> — J. William Thompson and Kim Sorvig, *Sustainable Landscape Construction*, 2008

What about Plastic Lumber?

Plastic or composite "lumber" is an increasingly popular material in landscape construction, for several reasons. Made partly or wholly from recycled plastic, it lasts longer than wood, produces no splinters, requires no painting, and costs less to maintain than wood. It is also is impervious to rot, termites, fire, and even vandalism (making it highly desirable for public outdoor furniture and other structures). Many consider this a green material, because it is seen as a good way to keep plastic waste out of landfills.

On the other hand, this material tends to have less structural strength than wood, and it can cost three times as much. Even worse, the "sawdust" produced when plastic lumber is worked is not only unusable for the purposes that natural sawdust could serve, but it contains polymer micro-beads that are harmful to animals and ecosystems. And finally, because plastic lumber is a hybrid of natural and industrial materials, it is neither compostable nor recyclable. All waste created during construction or later disassembly will end up in the landfill.

MATERIAL (**X-3**) **Metal**

Overview

The metal we use most often in our landscape projects is usually some version of steel. Wrought iron, once common in light-posts, fences, and ornamental gates, rarely suits our needs now. Copper, brass, and bronze might show up as fittings or fixtures, and aluminum plays an increasingly common role in outdoor furniture, planter boxes, and other small structures. But steel—strong and durable–is the metal that we use the most, for everything from garden edging to fences, benches, planters, and poles, from deck girders and posts to arbors, trellises, and gazebos. Given its widespread use, let's look at how climate-wise this versatile material is.

Durability

When properly protected from corrosion, steel can last for decades or even centuries, making it the most durable of all materials available to us now,

Credit: SReed

Figure X-8: Thick sheet metal, in this case allowed to rust, exemplifies a space-saving, low-maintenance, and climate-friendly approach to building low walls and beds in small landscapes.

Credit: SReed

Figure X-9: For this wheelchair ramp, traditional metal railings fit harmoniously with a "mixed-media" landscape of cut granite, modular bluestone pavers, and asphalt paving.

except stone. In some projects, letting steel corrode may be perfectly acceptable; the rust itself can produce an attractive patina on sheet steel used for low retaining walls or other containers. But in general it's preferable to prevent surface oxidation that will eventually eat away the material. Toward this end, we have two options.

- *Stainless steel*: treated with chromium, molybdenum, and/or nickel to prevent corrosion; comes in countless shapes and sizes; can be cast, forged, welded, or soldered; is very hard and strong, and totally resistant to rust even in salt spray; very expensive.
- *Coated steel*: "hot-dip galvanized" involves immersing the metal in a bath of molten zinc at 840°F; recognized by its soft gray rough surface; often plays a role in outdoor construction; less expensive than stainless.

Carbon Footprint

Steel production is one of the world's leading sources of greenhouse gases, accounting for 5–7% of the world's total GHG emissions. Overall, the process emits about 1.8 tons of CO_2 for every ton of steel produced. However the steel industry as a whole has increased energy efficiency by about 60% in the past 50 years.

Transportation and installation, which often require heavy vehicles and other equipment, also tend to have a large carbon footprint. But stainless steel requires very little maintenance, and coated steel can be equally trouble-free if properly manufactured and handled.

Potential To Be Recycled/Reused

Steel itself is infinitely recyclable. It can be continually processed and reused without loss of performance. About 95% of structural steel manufactured in the US is fully recycled from previously used steel products. For galvanized steel, the zinc coating is also fully recyclable, as well as being an abundant natural metal. And since both the steel and zinc are recycled at the end, this product has a very small carbon footprint beyond the production phase. In addition, the by-products of steel making are also valuable resources; blast furnace slag is being used by the cement industry to reduce CO_2 emissions (see X-4 on page 283).

Local Availability

The steel industry involves complex global trade deals, environmental regulations, and many other economic and social factors. Needless to say, the carbon footprint of steel could be reduced by increasing local production, and perhaps by exploring production methods that allow for smaller-scale operations. In the meantime, we can focus our efforts on using steel as efficiently as possible, reusing it whenever possible, and delivering it to local recycling operations whenever we are done with it.

Impact on Wildlife Habitat and Corridors

The mining of iron used in steel production disrupts land at every stage, from exploration to construction, operation, and closure of mines. Vegetation is cleared not just for the mine itself but also for roads, buildings, and power

Credit: SReed

Figure X-10: Gabions are metal mesh baskets or cages that hold chunks of local stone (or any other locally available material) and are stacked to create retaining walls and other structures. The result is a very small carbon footprint.

lines. Waste storage facilities and tailing ponds (where waste settles and the water is later recycled) expand over the life of the mine, extending habitat loss and deforestation well beyond the mine.

Mine operations can reduce their impact by salvaging rare or endangered species before operations begin, reducing the size of the mining area, and minimizing the amount of waste produced and stored. We can push for local regulations aimed at minimizing the worst impact.

Carbon Sequestration Potential

Steel production is a net source of CO_2 and other greenhouse gases. Steel's main advantage in this realm is that using steel can prevent the use of other

What about Gabions?

A hybrid of metal and stone, *gabions* are steel wire baskets tightly filled with cobbles, crushed rock, or other chunked material. The resulting large building blocks may be stacked in tiers or spread as a *revetment*, a sort of thick, sloped blanket, covering a vulnerable slope from tip to toe (this method has been especially useful in repairing/preventing beach erosion).

In recent years, gabions have morphed from unattractive but effective earth-grading solutions into attractive landscape features in their own right. They might still be used to retain a slope, or they might form the base of a freestanding wall, bench, outdoor grill, enclosure, or other landscape element. Instead of being basic woven wire, the baskets can now be stainless or coated, welded wire.

And instead of being filled with coarse jagged cobbles, their contents can run the gamut from multicolored glass rocks to recycled bottles, or even pine cones.

The big advantage of gabions is that an everyday homeowner can move and install them with ease, and fill them with a wide range of locally available decorative materials. They are also much more affordable than rock or concrete walls, but equally durable if the steel is coated and the structure is built properly, i.e., with a good base and sensible design. And, as with all steel, the metal itself is wholly recyclable. As such, gabions have one of the smallest climate footprints of all construction materials currently available.

materials with a larger carbon footprint, such as concrete, and can potentially reduce the felling of old-growth forests, and the associated loss of habitat and sequestered carbon. That said, the advantages of using steel have little to do with carbon sequestration.

Last Thoughts

Steel's complete recyclability, along with its extreme durability, strength, low maintenance, and versatility in reuse compensate somewhat for carbon emissions associated with its production. And as the industry continues to work toward greater energy-efficiency and more effective procedures for reducing greenhouse gas emissions, steel will likely become an even more climate-friendly material for our landscape projects.

MATERIAL X-4 Concrete

Figure X-11: Concrete's extreme durability is one of its greatest assets; in the right conditions it can even blend quite nicely into an established landscape.

Overview

Concrete production worldwide is a $100 billion industry. Twice as much concrete is used around the world than the total of all other building materials—including wood, steel, plastic, and aluminum. The only thing we use more of than concrete, ton for ton, is water.

Concrete is made by mixing cement (almost always Portland cement) with water and some kind of aggregate material such as sand, gravel, or crushed stone. The resulting pebbly or grainy soup gets poured into a form or mold. As the cement chemically reacts with the water, the paste hardens (cures) around the aggregate particles, producing a material that's almost as hard as rock.

Note: Although often used interchangeably, the word *cement* is not the same thing as *concrete*. Cement is the binder, a powdery

Credit: SReed

material that also forms the basis of mortar, grout, and stucco. Concrete is the result of mixing cement with water and other things.

Durability

Like rock, concrete is tough and strong. It can resist ultraviolet radiation, insect attack, and most chemical action. And it can withstand a lot of pressure. What it can't do at all, though, is bend. Concrete is very weak in situations that cause tension or stretching. As a result, concrete is often reinforced with embedded steel, in the form of either sheet mesh or parallel/interwoven rods, called *rebar*.

Concrete generally lasts about 30 to 50 years (although sometimes as long as 100 years). One main reason it deteriorates is the rusting of its reinforcing metals; this occurs within the concrete even without exposure to moisture or air. Other reasons concrete can break down include freeze-thaw action, the movement of frost-prone ground, penetration of water that freezes and thaws, and the effects of road salt. However, for most landscape construction projects, concrete remains one of the most durable materials available.

Figure X-12: The light color of these pre-cast concrete pavers makes them an ideal material for cool pavement. Note, however, that a climate-wise driveway would not be quite so expensive as this one, and would ideally include one or more shade trees.

Credit: SReed

Carbon Footprint

The concrete industry is one of the world's largest producers of carbon dioxide, generating 5–10% of man-made emissions worldwide. Most of this is the result of manufacturing Portland cement, which involves burning tremendous amounts of fuel to heat calcium carbonate up to about 2700°F (1500°C). Nearly one ton of CO_2 is emitted for every ton of cement produced.

Transporting the concrete batch to the job site and applying/spreading it requires heavy machinery, but in general the majority (about 70%) of concrete's carbon footprint is due to the

What about Asphalt?

Asphalt is a mixture of about 95% sand, gravel, and small stones, bound together by a thick tarry substance that is variously called *asphalt*, *asphalt cement*, *blacktop* or, most accurately, *bitumen*. Like concrete, asphalt's ingredients and mixing process vary widely with location and the intended use. Whatever the mix, asphalt is less durable than concrete, with a maximum lifespan of about 25 years. Many factors such as climate, traffic, preparation, and maintenance practices will shorten or lengthen that period.

The carbon footprint of asphalt is smaller than of concrete because it doesn't require energy-intensive Portland cement. Further, nearly all asphalt that gets removed from a road is either reused in new pavement or recycled as a base material for other driving surfaces. This complete reusability and small carbon footprint make it a good choice for paving small areas or oddly shaped terrain, especially in cold climates where its capacity to flex with ground movement is a big advantage.

cement manufacture itself. In recent years, the concrete industry has begun to find ways to help shrink concrete's carbon footprint by using industrial by products such as fly ash (from power plants) or slag (from steel mills) in place of mined aggregates.

Potential To Be Recycled/Reused

Recycling is an increasingly common choice when concrete structures are demolished or renovated. Rubble that in the past would have routinely been trucked to the landfill is now often collected and delivered to crushing operations. Any rebar in the waste is removed by magnets and recycled. Recycled concrete can be used as gravel in new construction, as base material for road pavement, for erosion control riprap, in gabions (See X-3 on page 279), or as aggregate for new concrete.

Local Availability

Concrete supplies are among the most widely available building material, with aggregate almost always supplied by local sources. In addition, many local contractors are usually available to do, or help with, the installation.

What about Concrete Blocks?

Often called *cinder blocks,* these familiar precast building blocks are made of Portland cement mixed with fine sand and gravel. For the purpose of landscape construction projects that don't need the strength of a building wall, the best choice may be *aerated blocks,* or *aircrete.* These blocks are made with industrial waste as a prime ingredient, in a process that forms air voids to produce a cellular structure, and consumes about 25% less energy than other concrete blocks. The lighter weight also saves on energy transportation. *Note:* The question of concrete *unit-pavers* vs. standard clay brick is addressed in X-5 on page 288.

Impact on Wildlife Habitat and Corridors

Concrete poses two kinds of problems for wildlife. First, cement production generates tremendous amounts of dust, and the mining of aggregate materials obliterates natural areas. Second, and perhaps more problematic, concrete structures and paved surfaces are completely inhospitable to wildlife. Multi-lane highways are particularly deadly: not only are animals killed while attempting to cross, but a highway itself can prevent movement between essential parts of some species' habitat, cutting off access to food and mating partners.

As the planet heats up and many animals are forced to move in search of more hospitable habitat, large expanses of concrete and urban development pose a tremendous barrier to their survival. (See more about wildlife corridors in VIII-6 on page 238.)

Carbon Sequestration Potential

Concrete overall is a carbon source, not a sink, due to the tremendous amount of CO_2 emitted during production of Portland cement.

Last Thoughts

Although concrete manufacture alone accounts for a large amount of human-generated CO_2 emissions worldwide, the material's strength, durability, and versatility make it a seemingly essential component of Western civilization.

What about Porous/Pervious Pavement?

Porous pavement is an extremely climate-wise material. By lowering urban heat, it can reduce the energy we spend on keeping cool. By absorbing rainwater, it reduces strain on storm drains, pipes, and sewer infrastructure, thereby saving the energy cost of repair and replacement. And by lowering the volume of polluted and/or hot water running off streets and sidewalks, porous pavement can reduce harm to downstream species and habitat. Porous pavement exists in three basic forms:

- *Porous concrete* (or *asphalt*) is paving material made with open pores that allow water to pass through into a *recharge bed*, a thick layer of crushed stone designed to absorb and temporarily hold that water before it passes into the soil below. This type of pavement works well in large expanses of pavement such as parking lots. As with so much in sustainable landscaping, the decision to use this material is a balancing act: its benefits must be weighed against the carbon footprint of its production.

- *Grass pavers* consist of a matrix of connected cells (made of concrete or plastic), which together provide a firm, drive-able surface, while the soil within each cell supports the growth of grass. They are very versatile in shape and application, generally most appropriate for surfaces that get occasional instead of regular traffic, more affordable than other hard paving, and usually chosen for aesthetic purposes. To succeed, grass pavers require a reasonable amount of good topsoil and regular watering.

- *Modular* or *unit pavers* are individual pieces of paving material (in the form of brick, stone, or precast concrete blocks/bricks) dry-laid on a base that allows water to percolate through. This solution is the easiest and simplest for the average homeowner to install, and ideal for small-scale projects like patios, courtyards, and driveways. *Note*: The type of material used to fill the cracks between the units determines overall permeability. For example, granular sand will allow more water to pass through than fine dust, and of course any use of mortar or cement-products will eliminate the surface's porosity.

Nonetheless, for landscaping projects we recommend keeping the use of new concrete to a minimum, creatively using recycled concrete, and encouraging producers to use fly ash in cement-production and recycled aggregate in the final concrete batches.

MATERIAL **X-5** **Earthen Materials**

Overview

A great benefit of using earthen materials such as brick, adobe, sand, gravel, and other earth-based products is that they generally have the smallest climate footprint of all materials available for landscape projects. Here's why:

- They are almost always locally available or locally made, so CO_2 emissions from transport are minimal.
- Because they are usually produced on a small scale, the extent of their impact on wildlife and natural areas tends to be much smaller than, for example, limestone quarries needed for cement production, massive logging operations that provide commercial lumber, iron-ore mines, and steel mills.
- Most of these materials are relatively low-tech in their installation and repair, and local contractors can install them easily, sometimes by hand.
- Most of them are completely recyclable.
- Most of them are also wholly or partially permeable, which helps cool the ground and sustain underground ecosystems.

Durability

The main disadvantage of earth-based materials is that, with a few exceptions, they tend to be less durable than stone, metal, or concrete. More vulnerable to gradual breakdown or erosion, these "recyclable" qualities might be considered advantages in one sense, but from the standpoint of energy (and CO_2) expended to replace or repair these materials, their relatively shorter lifetime is not wholly beneficial. Again, it's a question of balancing the benefits and drawbacks.

Below is a general description of several common materials' unique properties, uses, and benefits, in relation to their contribution to climate change and their carbon footprints.

Materials in block or modular form
Fired brick

- Made by hardening clay in kilns heated to about 2000°F (1100°C).
- This carbon footprint is moderate compared to concrete and asphalt.
- Innovative products replace some of the clay with recycled fly ash.
- Usually locally manufactured, with local raw materials.
- Useful for either pavement or walls/structures.
- Can have a long lifespan, as long as non-porous brick (specifically made for outdoor construction) is used.
- When used to build walls, the mortar (a cement product) increases the carbon footprint.
- High thermal mass; may be an advantage or drawback depending on location.
- Easy to use for paving small and/or hard-to-reach areas.
- More expensive, both to buy and install, than bulk pavement like concrete and asphalt.
- If permeability is desired, can be laid on a porous base and set so joints don't interlock tightly.

Credit: SReed

Figure X-13: Brick is a favored material for beautiful landscape construction that is also climate friendly due to its local availability and small carbon footprint.

Adobe

- Rectangular bricks made from sand, clay, and water mixed with fibrous or organic material such as straw, grass, twigs, or even horse or cow manures. The mixture is shaped into frames and left to dry in the sun.
- *Superadobe* is a type of earthbag construction that consists of long fabric tubes filled with a mixture like the one just described; the tubes flatten

Credit: SReed

Figure X-14: Adobe and stucco, shown in this New Mexico landscape, exemplify the ideal earth-friendly, low-impact, energy-efficient, ecologically harmonious, and climate-wise landscaping material.

Unit Pavers: Fired Clay Brick or Precast Concrete?

Fired clay bricks are tough and hard-wearing, with an anti-slip surface and non-porous exterior that requires no sealing. Their colors don't fade in UV light. In contrast, precast concrete pavers, require sealer to protect both the surface and the color, and they absorb more dirt and oils, making them higher-maintenance than brick. (They are also vulnerable to road salt.) This product is made in a wide range of colors and surface effects not available in traditional bricks. Despite their relatively larger carbon foot-print and other drawbacks compared to fired clay, precast products have become increasingly popular in recent years.

---

together when layered into a wall. This material is more likely to be used for house construction than landscape walls.

Compressed earth (or soil) block

- Made from soil, clay, and aggregate mechanically compressed in a form.
- May also be made with Portland cement, to create *compressed stabilized earth block* (CSEB).
- Different from mud bricks and adobe in that those are not compressed, but instead solidify through simple drying.

Alternatives to concrete blocks
Honeycomb clay block (Ziegel blocks)

- Formed with vertical perforations that reduce quantity of material, making it a lightweight alternative to concrete blocks, with a carbon footprint smaller than concrete.
- Recyclable, if dismantled carefully. Common in Europe for decades, usually used for home construction but equally applicable in landscape walls, borders, and other structures.

Hemp blocks

- A bio-composite made of the inner woody core of the hemp plant mixed with a lime-based binder.
- Often used for insulation, but also could be useful for outdoor free-

Credit: SReed

Figure X-15: In the end, materials gathered from one's own locale, such as crushed shells commonly used in coastal landscapes, may be the most climate-wise choice.

standing walls. Like all blocks, could be stuccoed with lime mortar for appearance.

- Very high thermal mass.
- Reusable and recyclable.
- Carbon negative as a result of CO_2 sequestration during photosynthesis.

Cob

Cob is a building material made from lumps of earth mixed with water and some sort of fibrous material (usually straw), and sometimes lime. It requires no special equipment and uses sustainable, renewable materials. Usually, it is used like adobe to create self-supporting walls, but unlike adobe, it is built wet.

What about Synthetic Materials?

The synthetic materials we might use in our landscape projects include products made from plastic, sponge rubber, and recycled rubber tires, along with plastic lumber (discussed in Section X-2). Although many synthetic products offer the benefits of low maintenance, durability, and the incorporation of recycled materials, the large carbon footprint and ecological impact of their production, coupled with their potential to be toxic to both humans and nature (especially true for *crumb rubber* often used in playgrounds), along with their inability to be further recycled, decrease their appeal as a climate-wise option.

It requires no forms, no ramming, cement, or rectilinear blocks. Like building with clay, cob lends itself to organic shapes.

Materials in loose or granular form
Stabilized soil

- Made by mixing dry Portland cement or some other binder (5–15% of the total) with native subsoil, then wetting the mixture and allowing it to cure.
- No aggregate or sand needed, which reduces cost and carbon footprint. Ideal for foot and bike traffic: paths, wheelchair trails, small patios, etc.
- Because earth-based materials tend to be less strong and durable than concrete and asphalt pavement, most are used only for light traffic areas such as paths, trails, and patios.

Sand and gravel

- These are usually the least expensive materials because they are used just as they are removed from the ground, with no added processing.
- Very small carbon footprint.
- Almost always porous to some degree.
- Easy to install, maintain and repair, useful for rustic paving, but will probably require more frequent maintenance.

Local crushed or pebbly materials

- Crushed sea-shells, crushed recycled brick, river rock, glacial outwash pebbles, decomposed granite, or other local angular and self-compacting stone fragments.
- All can serve well as surfacing material for light vehicle traffic (with an adequate base to prevent erosion and/or rutting) or for areas of foot traffic.
- All have a very small carbon footprint, with the limited CO_2 emissions due only to crushing operations and limited transport.

Last Thoughts

Using any of the earthen materials discussed here will give a landscape project the smallest possible climate footprint by minimizing CO_2 emissions and harm to the environment while maximizing the usefulness of locally available resources that are, for the most part, easily renewed, recycled, or reusable.

Conclusion

In this era of climate change, our landscapes can be so much more than beautiful havens and displays. Instead of just sitting there looking pretty, they can *do* something!

Every landscape can be a place where we shrink our carbon footprint and reduce greenhouse gas emissions.

Every property can grow more plants to help keep the planet cool and build carbon in the soil, where some of that carbon will be sequestered for the long term.

Every garden and yard can contain an abundance of native plants useful to pollinators, dragonflies, frogs, newts, and other wildlife that help keep nature in balance.

Credit: SReed

Climate-wise landscapes help cool the planet and provide rich habitat while also being comfortable, durable, livable and beautiful.

Every landscape can include diverse ecosystems that are hospitable to species—both plant and animal—that are stressed by climate change. And these places can be managed with a light touch to reduce our use of fossil fuels and leave some room for natural processes to work as they were designed to work.

Every yard is a place where we can express our dawning realization that we are the stewards of this planet. And who better to lead the way in taking care of this planet than the millions of gardeners, homeowners, landscapers, and property managers who already tend their own gardens and property?

In 2010, the conclusion to *Energy-Wise Landscape Design* said this: "Consider the unexpected beauty of meadow grasses and flowers rippling in a breeze. Or front yards full of fruit trees and vegetables. And what about a solar array as garden sculpture? It's time for us to imagine a new ideal, one in which beauty is not just a conventional norm or a familiar picture, but also an expression of our values. Now in the 21st century, we can shape our landscapes so that in addition to looking good they will also work for our own good and for the good of the larger world."

Nearly ten years later, with climate change now a certainty, that message seems more important than ever. Our gardens and yards are places where, with just a slight shift in attitude and some minor adjustments in our practices, we can be part of the remedy. Governments and corporations will do what they will. But we don't need to wait for them. Millions of us can take the countless small actions that will be needed to help curb and reverse climate change. We can do this, one climate-wise landscape at a time. This book has shown you how.

Endnotes

Introduction

1. US EPA. "Global Greenhouse Gas Emissions Data," epa.gov
2. "15 Sources of Greenhouse Gases," allianz.com
3. NASA. "Climate Change: How Do We Know?" nasa.gov
4. NOAA. "Global Climate Change Indicators," noaa.gov
5. IPCC. "Climate Change 2014 Synthesis Report Summary for Policymakers," ipcc.ch
6. National Academies Press. "Attribution Of Extreme Weather Events in the Context of Climate Change," 2016, nap.edu
7. Derksen, C. and R. Brown. "Spring Snow Cover Extent Reductions in the 2008–2012 Period Exceeding Climate Model Projections," *Geophysical Research Letters* 39, October 19, 2012.
8. NOAA "What is Ocean Acidification?" noaa.gov
9. Rattan, L. "Carbon Sequestration," 2008, accessed from ncbi.nlm.nih.gov
10. McKibben, B. *The End of Nature.* Random House, 2006, p 10.

Section I: Lawn

1. Webber, M.E. "Energy versus Water: Solving Both Crises Together," *Scientific American* September 1, 2008, scientificamerican.com
2. Milesi, E. et al. "A Strategy for Mapping and Modeling the Ecological Effects of U.S. Lawns," isprs.org
3. Seaman, G. "Lawn Care Chemicals: How Toxic Are They?" January 28, 2009, eartheasy.com
4. Malinoski, M.K., J. Traunfeld, and D. Clement, "IPM: A Common Sense Approach to Managing Problems in Your Landscape," (HG62), extension.umd.edu
5. LeCompte, C. "Fertilizer Plants Spring Up to Take Advantage of U.S.'s Cheap Natural Gas," *Scientific American* April 25, 2013, scientificamerican.com
6. Mole, B. "Fertilizer Produces Far More Greenhouse Gas Than Expected," *Science News* June 9, 2014, sciencenews.org

7. Philpott, T. "New Research: Synthetic Nitrogen Destroys Soil Carbon, Undermines Soil Health," February 24, 2010, grist.org

8. National Pesticide Information Center. "2,4-D," npic.orst.edu

9. Heimann, M.F., and R.C. Newman. "Plant Injury Due to Turfgrass Broadleaf Weed Herbicides," learningstore.uwex.edu

10. Becker, A. "How to Calculate the Carbon Footprint of Your Lawn Mower," April 25, 2017, sciencing.com

11. Townsend-Small, A., and Claudia I. Czimczik. "Carbon sequestration and greenhouse gas emissions in urban turf," *Geophysical Research Letters* January 22, 2010, onlinelibrary.wiley.com

12. US EPA. "Greenscaping: the Easy Way to a Greener, Healthier Yard," epa.gov

13. Becker, A. "How to Calculate the Carbon Footprint of Your Lawn Mower," April 25, 2017, sciencing.com

14. NH Dept. of Environmental Services. "Preventing Pollution in Your Own Backyard," des.nh.gov

Section II: Trees and Shrubs

1. Evans, E. "Tree Facts," North Carolina State University, projects.ncsu.edu

2. Perry, T.O. "Tree Roots: Facts and Fallacies," *Journal of Arboriculture* 8: 197–211, 1982.

3. Energy.gov. "Landscaping for Energy-Efficient Homes," energy.gov.

4. "Benefits of Trees," arborday.org

5. US EPA. "Using Trees and Vegetation to Reduce Heat Islands," epa.gov

6. Dr. E.G. McPherson, Center for Urban Forest Research.

7. "Benefits of Trees," arborday.org

Section III: Water

1. Copeland, C., and N. Carter, Congressional Research Service. "Energy-Water Nexus: The Water Sector's Energy," fas.org, January 24, 2017.

2. Geological Survey. "Transpiration: The Water Cycle," water.usgs.gov

3. Penick, P. *The Water-Saving Garden*. Ten Speed Press, 2016, 143–144.

Section IV: Ecosystems

1. National Wildlife Federation. "Effects on Wildlife and Habitat," nwf.org

Section V: Soil

1. Schwartz, J.D. "Soil as Carbon Storehouse: New Weapon in Climate Fight?" March 4, 2014, e360.yale.edu

2. Pimental, D. "Population Growth and the Environment: Planetary Stewardship," *Electronic Green Journal* 1(9) December, 1998.

3. Philpott, T. "New Research," Feb. 24, 2010, grist.org

4. Biello, D. "Peat and Repeat: Can Major Carbon Sinks Be Restored by Re-wetting the World's Drained Bogs?" *Scientific American* December 8, 2009, scientificamerican.com

5. Chalker-Scott, L. "The Myth of Paper-Based Sheet Mulch," Washington State University, https://s3.wp.wsu.edu

Section VIII: Urban Issues

1. Roa-Espinosa, A. et al. "Predicting the Impact of Urban Development on Stream Temperature Using a Thermal Urban Runoff Model (TURM)," National Conference on Urban Stormwater: Enhancing Programs at the Local Level. February 17–20, 2003, Chicago.

2. Kenward, A. et al. "Summer in the City: Hot and Getting Hotter," Climate Central, 2014, climatecentral.org

3. Huang, J., H. Akbari, and H. Taha. "The Wind-Shielding and Shading Effects of Trees on Residential Heating and Cooling Requirements," ASHRAE Winter Meeting, American Society of Heating, Refrigerating and Air-Conditioning Engineers, 1990, Atlanta, Georgia.

4. Akbari, H. et al. "Peak Power and Cooling Energy Savings of Shade Trees," *Energy and Buildings* 25:139–148, 1997.

5. Scott, K., J.R. Simpson, and E.G. McPherson. "Effects of Tree Cover on Parking Lot Microclimate and Vehicle Emissions," *Journal of Arboriculture* 25(3), 1999.

6. Kurn, D. et al. "The Potential for Reducing Urban Air Temperatures and Energy Consumption through Vegetative Cooling," ACEEE Summer Study on Energy Efficiency in Buildings, American Council for an Energy Efficient Economy. Pacific Grove, California, 1994.

7. Liu, K. and B. Bass. "Performance of Green Roof Systems," National Research Council Canada, Report No. NRCC-47705. Toronto, Canada, 2005.

8. Green Roofs for Healthy Cities. "Green Roof Industry Survey" July 13, 2017, greenroofs.org

9. Green Roofs for Healthy Cities. "Green Roof Industry Survey" July 13, 2017, greenroofs.org

10. US EPA. "Reducing Urban Heat Islands: Compendium of Strategies—Cool Roofs," epa.gov

11. Rose, L.S., H. Akbari, and H. Taha. 2003. Urban fabric analyses conducted by Lawrence Berkeley National Laboratory.

12. Pomerantz, M. et al. "The Effect of Pavements' Temperatures on Air Temperatures in Large Cities." Lawrence Berkeley National Laboratory, Berkeley, CA, 2000.

13. Hirtenstein, Anna. "Solar-Panel Roads to Be Built on Four Continents Next Year," Bloomberg, November 23, 2016, bloomberg.com

Section IX: Food

1. Cleveland, D.A. et al. "The Potential for Urban Household Vegetable Gardens to Reduce Greenhouse Gas Emissions," *Landscape and Urban Planning* 157: 365–374, January 2017, sciencedirect.com

Index

About the Authors

Sue Reed is a registered Landscape Architect, ecological site planner, native plant advocate, writer, and photographer. For the past 30 years, she has helped hundreds of homeowners across western New England create landscapes that are ecologically vibrant, comfortably livable, energy conserving, and beautiful.

After graduating from the University of Rochester in 1975, Sue became a master woodworker in Cambridge, Massachusetts, where she made modern hardwood furniture for five years and historic-reproduction French harpsichords for the next six. She earned her Master of Arts degree in 1987 from the Conway School of Landscape Design, and served as adjunct faculty there between 1991 and 2007, becoming licensed as a Landscape Architecture in 1994. Her one-woman design business, which started in 1991, continues to the present. Sue's award-winning first book, *Energy-Wise Landscape Design*, was published by New Society Publishers in 2010.

After receiving her M.S. in Botany from the University of Maryland, Ginny Stibolt took a 23-year journey into computer technology and website design. When she moved to north Florida in 2004, she thought she would finally have time to garden, but failed to account for its different climate. After teaching herself a whole new way to garden in this new world, she began writing a column called "Adventures of a Transplanted Gardener" to share her experiences. Soon, joining the Florida Native Plant Society changed her thinking from what *could* grow in Florida to what belonged there.

She has written four books in the last eight years, all published by the University Press of Florida. They are: *Sustainable Gardening for Florida* (2009); *Organic Methods for Vegetable Gardening in Florida* (2013); *The Art of Maintaining a Florida Native Landscape* (2015); and *Step-by-Step to a Florida Native Yard* (2018). She has also authored hundreds of articles both online and in print. Links to all of them may be found at Greengardeningmatters.com.

ABOUT NEW SOCIETY PUBLISHERS

New Society Publishers is an activist, solutions-oriented publisher focused on publishing books for a world of change. Our books offer tips, tools, and insights from leading experts in sustainable building, homesteading, climate change, environment, conscientious commerce, renewable energy, and more—positive solutions for troubled times.

We're proud to hold to the highest environmental and social standards of any publisher in North America. This is why some of our books might cost a little more. We think it's worth it!

- We print all our books in North America, never overseas

- All our books are printed on 100% **post-consumer recycled paper**, processed chlorine-free, with low-VOC vegetable-based inks (since 2002)

- Our corporate structure is an innovative employee shareholder agreement, so we're one-third employee-owned (since 2015)

- We're carbon-neutral (since 2006)

- We're certified as a B Corporation (since 2016)

At New Society Publishers, we care deeply about *what* we publish—but also about *how* we do business.

New Society Publishers
ENVIRONMENTAL BENEFITS STATEMENT

For every 5,000 books printed, New Society saves the following resources:[1]

41	Trees
3,728	Pounds of Solid Waste
4,102	Gallons of Water
5,350	Kilowatt Hours of Electricity
6,777	Pounds of Greenhouse Gases
29	Pounds of HAPs, VOCs, and AOX Combined
10	Cubic Yards of Landfill Space

[1] Environmental benefits are calculated based on research done by the Environmental Defense Fund and other members of the Paper Task Force who study the environmental impacts of the paper industry.